T0213982

Blocks, Towards Energy-efficient, Coarse-grained Reconfigurable Architectures

Mark Wijtvliet • Henk Corporaal • Akash Kumar

Blocks, Towards Energy-efficient, Coarse-grained Reconfigurable Architectures

 Springer

Mark Wijtvliet
TU Dresden
Dresden, Germany

Akash Kumar
TU Dresden
Dresden, Germany

Henk Corporaal
Eindhoven University of Technology
Eindhoven, The Netherlands

ISBN 978-3-030-79776-8   ISBN 978-3-030-79774-4   (eBook)
https://doi.org/10.1007/978-3-030-79774-4

This Springer imprint is published by the registered company Springer Nature Switzerland AG
The registered company address is: Gewerbestrasse 11, 6330 Cham, Switzerland

Contents

List of Acronyms

Chapter 1
Introduction

Embedded processors can be found in almost any device that uses electricity. In devices like smart-phones and TVs, the presence of a processor can easily be recognized, but in many cases this is not so clear. Examples thereof are devices like kitchen appliances, the thermostat on the wall, and even bank cards. In fact, already in 1999 close to 100% of microprocessor production went to embedded applications [1]. The other part, less than half a percent, goes to desktop computers, laptops, and notebooks.

In many cases, embedded processors are optimized for certain application domains. The most common application domains for embedded microprocessors are control applications and digital signal processing, or a mix thereof. Control applications can be tasks such as controlling a microwave oven or the cruise control of a car. Digital signal processing can be found in, for example, video processing on a mobile phone or audio processing in an MP3 player.

There are many situations where the energy efficiency of a microprocessor is crucial. This can be the case when such processors are used in battery powered or energy harvesting devices, but also in high performance compute clusters where energy usage is responsible for a large part of the operating cost of a data center. If the production volume is large enough, very high energy efficiency can be achieved by designing Application specific integrated circuits (ASICs). However, since the ASIC design is such an expensive process, it is more common to use off-the-shelf processors. These processors are, although optimized towards their target application domain, capable of executing many different types of applications. The downside is that this goes at the expense of energy efficiency and often also at the expense of performance.

The ideal situation would be to have a processor platform that approaches the energy efficiency of an ASIC but has the flexibility of a general purpose microprocessor to allow execution of many application types. This book introduces Blocks, an architecture that takes a step in this direction. It is a reconfigurable processor that can adapt at run-time to the application it needs to execute. Blocks

© The Author(s), under exclusive license to Springer Nature Switzerland AG 2022
M. Wijtvliet et al., *Blocks, Towards Energy-efficient, Coarse-grained Reconfigurable Architectures*, https://doi.org/10.1007/978-3-030-79774-4_1

is a Coarse-grained reconfigurable architecture (CGRA) that distinguishes itself by separating the data-path from the control-path by using two circuit switched networks. This allows for very energy-efficient execution of a wide range of applications with scalable performance.

This chapter introduces the reasoning behind the development of Blocks and a quick introduction into the field of energy-efficient (reconfigurable) computing. The structure of this chapter is as follows. Section 1.1 provides a brief introduction into the current processor landscape of reconfigurable and non-reconfigurable processor architectures. The section also introduces the design requirements for modern embedded microprocessors. Section 1.2 details the problem statement of this book; it shows that none of these architectures can fulfil all of these requirements, leaving space for a new architecture to be introduced. The Blocks architecture is proposed as a solution to this problem in Sect. 1.3. This section describes how Blocks is expected to meet all requirements simultaneously. Section 1.4 outlines the structure of this book. Finally, Sect. 1.5 concludes the content of this chapter.

The content of this book is based on [2].

1.1   The Embedded Processor Landscape

Many types of microprocessors have been developed throughout the past decades. Some computer architectures aim to provide support for a wide range of applications, while others focus on a smaller set of applications. Specialized architectures perform better in some aspect within their application domain, e.g., performance or energy efficiency. A specialized architecture can also be a reconfigurable fabric that can adapt to the application set. In this section, the more common processor and reconfigurable fabric types that can be found in embedded computer systems are introduced and compared in terms of their properties. Flynn's taxonomy [3] provides some guidelines on how to do this. In this taxonomy, shown in Table 1.1, data and instructions are considered for both a single element and multiple elements at the same time. For example, a processor operating on one instruction and multiple data elements at a time is classified as an Single instruction multiple data (SIMD) processor. Similarly, this can be done for the other three combinations. Although an Multiple instruction single data (MISD) processor is technically possible, opinions on which architectures classify as an MISD processor is open to debate. MISD processors may be a good choice for applications where reliable execution is vital. In this case multiple redundant computations are performed on the same data

Table 1.1 Flynn's taxonomy of computer architectures

		Data	
		Single	Multiple
Instructions	Single	SISD	SIMD
	Multiple	MISD	MIMD

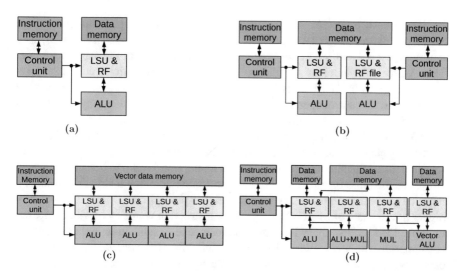

Fig. 1.1 Four different processor architecture types. Each processor type contains one or more control units that issue instructions to function units in the architecture. The processors shown contain a load–store unit (LSU), a register file, an arithmetic logic unit (ALU), and, optionally, a multiplier (MUL). (**a**) A general purpose processor has a control unit, a register file, and an ALU. The control unit controls only one ALU at a time. (**b**) A multi-core or many-core contains multiple independent control units and data-paths. These are connected via either a shared memory or an interconnect network. (**c**) An SIMD processor has a single control unit that controls the operations of several identical vector lanes. Each vector lane performs the same instruction, but on different data. (**d**) A VLIW has a single decoder, but unlike an SIMD each issue slot in the architecture can perform another operation. The data-path of a VLIW can be more irregular than that of an SIMD. The function units (including LSUs) are arranged into issue slots

1.1.1   Processors

The most common embedded processors are General purpose processors (GPPs), and these are not optimized for any specific application and contain instructions that most applications typically require. Examples of these embedded processors are the ARM Cortex series and the RISC-V processors. Figure 1.1a shows a schematic example of a GPP. Due to their general applicability, performance is moderate and energy consumption is relatively high. Often, GPPs are extended with accelerators, special hardware additions that optimize the processor for certain often recurring tasks. These accelerators improve performance and energy efficiency for some applications. Many-core processors are recently gaining attention in the embedded processor world. Although they have been researched in the past in the context of embedded systems, they are more common in high performance computing applications. Many-core processors typically contain a dynamically routed interconnect, like a bus or a network-on-chip, that allows fast data transfer between light-weight GPPs connected to the interconnect. Figure 1.1b shows a two-core processor that uses a shared interconnect for communication.

Examples of processor that are more aimed towards an application domain are SIMD processors and VLIW processors. SIMD processors, shown in Fig. 1.1c, have multiple vector lanes where all processing elements perform the same operation. This reduces the number of fetched and decoded instructions and, therefore, some energy overhead. However, they can only efficiently execute data-parallel applications. To control operations of the vector processor, SIMDs may be extended with a dedicated scalar control processor. Very long instruction word processors (VLIWs) on the other hand are quite the opposite of SIMD; they can control each processing element (an issue slot) individually, as shown in Fig. 1.1d. This allows for efficient application mapping as long as instruction-level parallelism is present. However, the wide instruction memories, and more complex instruction decoders, required to control the processor, lead to a higher power draw. Despite the typically higher power draw of SIMDs and Very long instruction words (VLIWs), compared to embedded GPPs, their increased performance results in a reduced energy consumption for applications that map efficiently. However, when applications cannot be efficiently mapped, the energy efficiency of these processors can decrease significantly. This is the case when, for example, an SIMD processor has to execute an application which does not contain much Data-level parallelism (DLP).

A variant on SIMDs can be found in Graphics processor units (GPUs). Although originally intended for graphics processing, these architectures are now very popular for high performance computing. A GPU performs SIMD but allows fast context switching between threads and supports tracking of threads that diverge (e.g., at branches). This execution model is generally called Single instruction multiple thread (SIMT). The advantage of SIMT is better hardware utilization and latency hiding when accessing memories. However, the increased architecture complexity comes at the cost of higher power draw.

The processor architectures described in this section can be classified according to Flynn's taxonomy, as shown in Table 1.2. More specialized are Application specific processors (ASPs), and these processors are specially designed for a set of applications that is targeted to a specific application domain, sometimes even

Table 1.2 Processor architectures and their execution models

Architecture	Description	Execution model
GPP	Typically single instruction and data, but can have SIMD extensions.	SISD (SIMD)
Many-core	Multiple independent processors on multiple data elements.	MIMD
Vector architecture	A single instruction controlling multiple parallel vector lanes.	SIMD
GPU	Multiple vector units	MIMD
VLIW	Multiple issue slots with their own operations operating on multiple data elements.	SI(MO)SD[a]

[a]A VLIW contains a single very wide instruction that controls multiple operations (in issue slots). These issue slots typically operate on a single data element

a single application. They contain only the functionality, and internal connections, that the application(s) requires. This leads to high performance and a very good energy efficiency, but at the expense of flexibility. For many applications, it is not feasible to design an ASP and manufacture it on ASIC technology due to their high design and production cost, unless the production volume is very high.

1.1.2 Reconfigurable Logic

The best known type of fine-grained reconfigurable logic devices is the Field programmable gate arrays (FPGAs). These devices can be seen as a 'box' of fine-grained logic elements that can be connected together in order to implement almost any digital electronic circuits, including complete processors. Logic functions are implemented using Look-up tables (LUTs) that are configured by programming a configuration memory and are part of a Configurable logic block (CLB). The number of input bits of a CLB can vary but typically ranges from 4 to 6 inputs. CLBs with more than four inputs are often internally split into two smaller LUTs to reduce the number of configuration memory entries. For example, a 6-input CLB may be split into two four-input LUTs, where some of the inputs are reused between LUTs. Figure 1.2 shows an example of such a CLB. The LUTs are connected via switch-boxes, which provide circuit switched (unbuffered) connections between LUTs. These switch-boxes are also configured via a configuration memory. FPGAs are very popular for prototyping but can also be found in production devices. Their main advantage over ASICs is their reconfigurability. However, this results in a high penalty in performance, energy efficiency, and chip area.

For many applications bit-level reconfiguration granularity is not required, and therefore fine-grained implementation is too costly. Most digital signal processing applications operate on fixed point or floating point values that are multiple bits wide. This leads to the development of CGRAs. There exist many varieties of CGRAs. Some CGRAs have a circuit switched network, like FPGAs, while others have dynamically routed networks. The granularity of the compute elements,

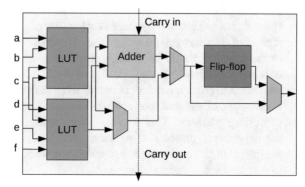

Fig. 1.2 An example of a CLB as used in FPGAs. The LUTs implement the designed logic, assisted by adders and small memory element (flip-flops). The multiplexers are configured by the FPGA configuration to implement the designed digital logic. A carry-chain allows the adders of multiple CLBs to be combined into wider adders

function units (FUs), also varies widely over the different architectures. In general, CGRAs have Functional units (FUs) of 8- to 32-bit wide that can be connected over the interconnect. These function units are either static or have a local instruction sequencer that allows temporal reuse of the FU hardware. Temporal reuse means that hardware can be reused for multiple (different) operations that occur in different clock cycles. For example, an Arithmetic logic unit (ALU) may first be used for an addition, followed by a logic shift in the next clock cycle. CGRAs are somewhat less flexible than FPGAs, and they cannot efficiently implement any digital circuit. However, they can be much more energy efficient than FPGAs for many signal processing and control applications where bit-level granularity is not required. Despite this, surprisingly, most CGRAs in the past have focused on performance rather than energy efficiency. Energy efficiency is the amount of energy consumed per executed operation, or, inversely, how much work can be performed for a certain amount of energy.

1.1.3 Processor Metrics

There are many architecture properties for which the suitability of processors and reconfigurable logic, for a specific application, can be measured and evaluated. The most common of these metrics are described below.

- **Performance**: for many embedded applications, there are latency or throughput constraints. For video filtering, for example, a new frame is expected to be produced at a predetermined rate (the throughput) to not interrupt video playback, while a control application expects a result within a specified amount of time (latency). Given the computational demands that the application has for computing each frame, the required processor performance can be determined.
- **Power**: the power draw of a processor during program execution is important for applications where a limited power budget is available. This can be the case in, for example, solar powered devices and energy harvesting sensors. Power limits can also be imposed due to thermal considerations, to prevent devices from becoming too hot. This can be very important for smart-phones and laptops, but also in data centres where active cooling of hardware increases costs and takes space that could be used by computer hardware.
- **Area**: the chip area that an architecture occupies. Chip area is directly connected to the price of a chip in volume production, not only because of the area it occupies on a wafer but also because of yield. A larger architecture has a higher likelihood to contain a defect in one of the transistors or wires on the die.
- **Flexibility**: often, multiple applications must be able to be executed on the processing platform. This requires some level of flexibility, which comes at the expense of power and performance. The goal is to choose a processor flexible enough for the application domain without compromising (too much) in performance or energy efficiency.

- **NRE Cost**: many embedded microprocessors are in devices that are developed for a low production cost. It is, therefore, important that the microprocessor does not contribute too significantly to the total production cost of the device. For microprocessors and other electronics on a chip, this is usually achieved by a high production volume. Cost can be divided into Non-recurrent engineering (NRE) cost and production cost. The first is cost for, e.g., ASIC design and manufacturing set-up, and the second is silicon area and packaging cost.

By themselves the above metrics do not provide direct insight into whether one processor is more suitable for an application than another, with the exception of performance comparisons. Multi-dimensional comparisons are required to select a processor that is expected to be energy efficient. To do so, more precise metrics are defined by combining those above. Some examples thereof are as follows:

- **Energy efficiency**: this metric indicates how efficiently a processor uses its energy supply. A processor, such as an SIMD, may have a higher power draw compared to a GPP, but it performs multiple operations per cycle. As long as the power scales slower than the number of operations performed, the energy efficiency will improve. This implies that even processors with a very high power draw can be very energy efficient for a certain application.
- **Area efficiency**: it indicates how efficient the chip area of an architecture is used for computation. An architecture can have a very large area that can be efficiently used, like in an FPGA, or a very large area of which only a limited region can be used, like Big.Little architectures. The area efficiency metric is expressed as performance per area.

The properties of the processors, described in the sections before, for the embedded control and signal processing domains are shown in Table 1.3. In this table, cost is considered to be the cost to the designer of the end-product. For example, developing a product that uses a GPP hardly requires any ASIC NRE cost, while an ASP would.

Embedded GPPs are not optimized towards a specific application. Although this results in a flexible processor, the consequence is that the performance is not

Table 1.3 Architectural metrics. The classification ranges from '−−' to '++', where '−−' indicates very poor performance for a certain aspect and '++' very good performance

Processor type	Energy efficiency	Area efficiency	Flexibility	Cost
GPP	−	−	+	++
Many-core	o	−	−	+
SIMD	+	+	o	+
GPU	o	o	+	o
VLIW	o	o	+	+
ASP	++	++	−−	−−
FPGA	+	o	++	o
CGRA	++	+	++	+

that good. However, since embedded microprocessors tend to be very simple, their power draw is quite acceptable. Since the energy efficiency depends on both the performance and the power draw, performance is typically not very high. GPPs are available off-the-shelf and are produced in very high volumes, making them a very cheap option. Some GPPs include limited vector widths of SIMD support to increase energy efficiency somewhat. Many-core processors are not that common yet in embedded applications, thereby increasing their cost as their production volume is relatively small. However, multi-core becomes more popular; an example thereof is the ARM Big.Little architecture. The flexibility of many-core processors is similar to GPPs, but their energy efficiency is generally somewhat better due to data reuse and amortization of overhead. However, since there are more (independent) cores available, performance is better than GPPs. SIMDs have higher energy and area efficiency due to reuse of the instruction decoder hardware over all vector lanes. However, since all units now have to perform the same operation, this comes at expense of some flexibility. This is not the case for VLIWs as these can perform different operations for each issue slot, but this leads to more complex instruction decoders and memories and, therefore, lower energy and area efficiency. Both SIMD and VLIW processors are commercially available. Although not as common as GPPs, their off-the-shelf cost is lower than most many-core processors. ASPs perform very well with respect to performance and power aspects as they are optimized for the target application. At the same time, this makes them very inflexible; most other applications will suffer significant performance penalties on the same platform. Since ASPs are application specific, dedicated silicon designs have to be taped-out for each design, making ASPs very expensive.

FPGAs are some of the most flexible devices around, since they can be used to implement and prototype any (digital) circuit, including processors. However, due to the complex interconnect and logic block configuration, power draw is quite significant. Additionally, performance suffers due to the long critical paths. This leads to both a lower energy and area efficiency. Additionally, FPGAs have a comparatively large chip area, leading to a lower yield, and these devices are more expensive than most other processing platforms. CGRAs on the other hand have a simpler interconnect, leading to higher energy and area efficiency than FPGAs as well as better performance. This can be attributed to the fact that CGRAs usually use bus-based interconnects instead of individual wires. When CGRAs would be produced in similar volumes as FPGAs, their cost would be lower due to the reduced chip area. For many target domains for embedded control and signal processing applications, CGRAs can be considered to be equally flexible as FPGAs, since these domains often use fixed point arithmetic instead of bit-level operations. This can be observed by the increasing number of CGRA overlay frameworks on top of FPGAs, such as QuickDough [4] and a technique based on SCGRAs [5].

1.2   Processor Requirements for Embedded Processing

For embedded applications, it is important that certain latency or throughput requirements can be met. These constraints can be imposed by hard deadlines, e.g., in a car where the embedded processor actuates the brakes. This is an example of a latency requirement; if there is too much time between pressing the brake pedal and the actual actuation of the brakes, this may lead to dangerous situations. Some constraints can be based on soft deadlines; violating these deadlines does not cause potentially dangerous situations but may degrade the perceived quality experienced by the user. An example of such a situation is a missed frame in a video played on a mobile device. Most of these situations can be avoided when the embedded platform has scalable processing capabilities. This means that the available processing power can be adapted to the throughput or latency constraints of an application, or, specified as a requirement:

1. The processing platform shall provide scalable performance, allowing the architecture to adapt to the throughput and latency requirements of an application.

This requirement will be checked by evaluating the architecture that is introduced in this book on a set of benchmark applications. These applications are chosen such that they represent much of the work typically performed by digital signal processors such as SIMDs and VLIWs. The proposed architecture achieves the above requirement when it approaches, or ideally exceeds, performance achieved by both the SIMD and VLIW processors.

Modern embedded systems are increasingly battery powered; this stems from the desire to make applications more mobile. This holds not only for communication devices like mobile phones but also for, for example, medical equipment. An increasing array of body sensors is in development that can monitor health during normal daily activities. Such body sensors can not only be consumer applications such as heart rate monitoring watches but can also be medical implants that monitor brain activity and warn for oncoming seizures. For all these devices, battery lifetime is very important, either from a usability perspective or from a marketing point. To achieve good energy efficiency, the processor should provide a good balance between performance and power draw. For an energy constrained embedded device, the following requirement can be specified:

2. The processing platform shall provide high energy efficiency.

This requirement can be fulfilled by using an ASP, and these devices achieve very good energy efficiency by providing both high performance and low power for specific applications. However, it is quite common for embedded devices to outlive original algorithms and standards. An example of this is a phone with 4G support that cannot connect to a 5G network, even though they operate in the same radio frequency bands. This is, in part, due to the on-board hardware accelerators to process data for these standards but that cannot be updated to support these new standards. These accelerators often include ASPs which provide enough flexibility for post production bug fixes, but not for application updates. For the brain monitoring case described earlier, architectural flexibility is even more important, since algorithms change fast and can even be patient dependent. This requirement can be considered as achieved when the energy to execute the set of benchmark applications is lower than the energy required for fixed processors such as a GPP, an SIMD, and a VLIW. Additionally, the architecture introduced in this book must provide better energy and area efficiency than standard CGRAs with local instruction decoding.

Due to the high NRE cost of ASP production, it is usually not feasible to develop an ASP for each new application and certainly not for every individual patient. Flexibility therefore implies a reduction in cost, since the processor design is not bound to a specific application and can be reused. For a generally applicable embedded processing platform, a new requirement can, therefore, be introduced:

3. The processing platform shall be flexible [6], in order to allow new (high volume) applications or updates.

Ideally, a designer would like to use a processor architecture that fulfils all three requirements. However, Table 1.3 shows that there is currently no architecture that can provide this. Therefore, a trade-off has to be made. Unfortunately, this is not always straightforward. For example, a GPP scores quite well on flexibility but relatively poor on performance. Similarly, an FPGA provides good performance and flexibility, but power is rather high.

Although not perfect, CGRAs are a step in the right direction. They provide good performance and are quite flexible. Their flexibility allows for cost reduction by large volume production, since the same design can be reused in many applications. For Digital signal processing (DSP) applications, the main drawback of current CGRAs is that their energy efficiency is still relatively low, although better than FPGAs. If their power can be reduced, such that energy efficiency comes closer to ASPs, CGRAs can become a very interesting platform for future energy-efficient applications. This book focuses on just that a new type of CGRA that provides a flexible scalable performance *and* energy-efficient processing platform. Furthermore, this book will demonstrate that it is possible to outperform non-reconfigurable architectures in terms of performance and energy efficiency.

1.3   Architecture Proposal and Focus

This book introduces a novel CGRA, called Blocks. The Blocks architecture consists of an array of Functional units that are interconnected by a circuit switched network, similar to FPGAs. FUs in Blocks are heterogeneous which has several advantages. Firstly, the instruction width per FU can be reduced, which reduces instruction memory size and, therefore, power. Secondly, the function units themselves are simpler and smaller, compared to those in GPPs, which reduces power due to fewer toggling signals and area, respectively. The interconnect allows

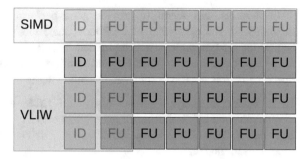

Fig. 1.3 An abstract representation of the Blocks architecture. A grid of function units (FUs) is configured to be controlled by instruction decoders (IDs). This allows configuration of SIMDs, VLIWs, specialized data-paths, and combinations thereof

the output of an FU to be arbitrarily connected to inputs of other FUs. Doing so allows the creation of application specific bypass networks, at run-time. Compared to an ASP, this will introduce some overhead, as there are now switch-boxes in the signal path, but flexibility is significantly improved. Due to the reconfigurable connections, Blocks can implement specialized data-paths, leading to very high performance. Therefore, Blocks is expected to fulfil the first requirement (Fig. 1.3).

In addition, Blocks has a secondary network that allows connecting Instruction decoders (IDs) to FUs. This sets Blocks apart from other reconfigurable architectures that perform decoding and instruction sequencing locally in the FU. Implementing the IDs as a separate function unit and allowing connection over a reconfigurable interconnect allow Blocks to perform SIMD style processing with scalable vector widths. This can be achieved by connecting a single ID to more than one FU. This causes all FUs to receive the same instruction and, therefore, perform the same operation. This effectively results in a vector processor. Doing so reduces the number of instruction decoders and instruction memories and, therefore, results in a reduction in power. Of course, there is some overhead from the reconfigurable network, but the overhead is smaller than individual memories and decoders. Combined with heterogeneous FUs and the choice for circuit switched networks, Blocks significantly reduces energy overhead for CGRAs. Blocks is, therefore, expected to provide high energy efficiency, thereby fulfilling the second requirement.

By employing multiple IDs that are connected to the same program counter, VLIW style processors can be implemented. Any combination of VLIW and SIMD is also possible, as are application specific structures. Blocks, therefore, allows any arbitrary processor structure to be instantiated on the reconfigurable fabric. These processor structures can be reconfigured for each application that requires execution on the platform, and this occurs at run-time. Blocks, therefore, allows very efficient application mapping for a wide variety of applications with little overhead in performance or energy. Furthermore, the structure of Blocks allows multi-processor structures to be implemented on the fabric, thus allowing task-level parallelism. Since Blocks is not optimized for a specific application or application domain but still performs well for many applications, it can be considered very flexible and therefore expected to satisfy the third requirement.

In short, Blocks seems to provide good performance, energy efficiency, and flexibility. In the remainder of this book, the Blocks architecture and reasoning behind the design choices will be treated. Furthermore, Blocks will be evaluated for a set of benchmark applications and additional improvements will be discussed. Specifically, this book will focus on the following topics:

1. An extensive analysis on the history of CGRAs, identifying their existing strengths and opportunities for energy reduction.
2. A new CGRA, Blocks. An introduction in its concept and impact on various execution models.
3. An extensive evaluation of Blocks, demonstrating that Blocks outperforms existing processor architectures.
4. An architectural model that enables fast exploration of the architecture in energy and area.

1.4 Book Outline

The remainder of this book is organized as follows:

Chapter 2 'CGRA Background and Related Work' presents an extensive investigation into existing coarse-grained reconfigurable architectures. This chapter also presents a definition for CGRAs and classifies existing architectures accordingly. This chapter finishes with an overview of possible research topics and guidelines for future CGRA development. This chapter is based on [7] and [8].

Chapter 3 'Concept of the Blocks Architecture' introduces the concept of the Blocks architecture and how it can be used to construct energy-efficient processors for various applications. Furthermore, this chapter describes the building blocks that make up the architecture. Lastly, an outlook on how multi-granular and approximate function units could be used in the Blocks framework is discussed. This chapter is based on [9].

Chapter 4 'Template and Tools' describes the design flow that makes up the software environment around Blocks. Design parameters and templates that are used to configure hardware generation are described. Subsequently, the process of hardware generation and executable generation is detailed.

Chapter 5 'Evaluation' provides an extensive evaluation on the performance, energy, and area properties of Blocks for various benchmark kernels. The experimental set-up is described, followed by an analysis of the results. In this chapter, Blocks is evaluated against a traditional CGRA, a commercial microprocessor, a VLIW, and an SIMD. This chapter is based on [9] and [10].

Chapter 6 'Architectural Model' describes how micro-benchmarks are used to profile the energy and area properties of the individual building Blocks that make up the Blocks architecture. These properties are used to compose energy and area models that can be used as early estimators. Such estimations are vital for design space exploration tools. The chapter finishes with an evaluation on the accuracy and performance of the model. This chapter is based on [11].

Chapter 7 'Case Study: The BrainSense Platform' presents a case study in which the Blocks architecture is used as part of a system-on-chip to implement an energy-efficient design for brain–computer interfacing. Using the architectural model from Chap. 6, properties of this architecture are estimated.

Chapter 8 'Conclusions and Future Work' finalizes this book by reviewing the conclusions and presenting future work.

1.5 Conclusions

The number of computing devices embedded in electronic products is ever increasing. Virtually every battery powered device now contains some form of on-board processing. For large production volumes, ASICs can be used to attain very high levels of energy efficiency. However, due to a high NRE, these cannot be designed for every product. At the same time, general purpose processors do not provide a sufficiently high energy efficiency to allow for long battery operation. A more energy-efficient yet programmable architecture can alleviate this problem.

Several processor types are discussed in this chapter, from general purpose processors to VLIWs. Some of these processor types are aimed towards a certain type of algorithms, making them less generally applicable. A reconfigurable architecture can overcome this by adapting the hardware to the algorithm. FPGAs can do so but are too fine grained and have large energy overhead. CGRAs provide more coarse-grained hardware blocks that are more suitable to signal processing applications. However, there is still a gap between the energy efficiency of ASICs and CGRAs.

A new CGRA architecture is proposed, Blocks, that aims to improve energy efficiency by separating the control and data networks. With Blocks, it is possible to construct processors that are closely matched to the application they are required to execute. For each application, Blocks can be reconfigured. The structure of the Blocks fabric allows various types of parallelism, including SIMD, to be efficiently supported. This is expected to provide improvements in energy efficiency.

References

1. J. Turley, *Embedded Processors by the Numbers*. https://www.eetimes.com/author.asp?section_id=36&doc_id=1287712 (1999) [Online; accessed 29-August-2018]
2. M. Wijtvliet, Blocks, a reconfigurable architecture combining energy efficiency and flexibility. PhD thesis. Eindhoven University of Technology, 2020
3. M.J. Flynn, Some computer organizations and their effectiveness. IEEE Trans. Comput. C-**21**(9), 948–960 (1972). ISSN: 0018-9340. https://doi.org/10.1109/TC.1972.5009071
4. C. Liu, H. Ng, H.K. So, QuickDough: a rapid FPGA loop accelerator design framework using soft CGRA overlay, in *2015 International Conference on Field Programmable Technology (FPT)*, December 2015, pp. 56–63. https://doi.org/10.1109/FPT.2015.7393130

5. C. Liu, H.-C. Ng, H.K.-H. So, Automatic nested loop acceleration on FPGAs using soft CGRA overlay. CoRR:abs/1509.00042 (2015). http://arxiv.org/abs/1509.00042
6. C.H. Van Berkel, Multi-core for mobile phones, in *Design, Automation Test in Europe Conference Exhibition, 2009. DATE '09* (2009), pp. 1260–1265. https://doi.org/10.1109/DATE.2009.5090858
7. M. Wijtvliet et al., Position paper: reaching intrinsic compute efficiency requires adaptable micro-architectures, in *Ninth International Workshop on Programmability and Architectures for Heterogeneous Multicores, MULTIPROG 2016* (2016)
8. M. Wijtvliet, L. Waeijen, H. Corporaal, Coarse grained reconfigurable architectures in the past 25 years: overview and classification, in *2016 International Conference on Embedded Computer Systems: Architectures, Modeling and Simulation (SAMOS)* (IEEE, New York, 2016), pp. 235–244
9. M. Wijtvliet et al., Blocks: redesigning coarse grained reconfigurable architectures for energy efficiency, in *2019 29th International Conference on Field Programmable Logic and Applications (FPL)*, September 2019, pp. 17–23. https://doi.org/10.1109/FPL.2019.00013
10. M. Wijtvliet, H. Corporaal, Blocks: challenging SIMDs and VLIWs with a reconfigurable architecture (2020, submitted)
11. M. Wijtvliet, H. Corporaal, A. Kumar, CGRA-EAM - Rapid energy and area estimation for coarse-grained reconfigurable architectures, in *TRETS 2021* (accepted, to appear). https://doi.org/10.1145/3468874

Chapter 2
CGRA Background and Related Work

Modern embedded applications require a high computational performance under severe energy constraints. In Chap. 1 some reasons for using CGRAs were given; in this chapter some more detailed examples will be given to show how CGRAs can contribute to increase energy efficiency. Mobile phones, for example, have to implement the 4G protocol, which has a workload of about 1000 GOPS [1]. Due to battery capacity limitations, the computation on a mobile phone has a processing budget of about 1 W. Thus, under these requirements, each computational operation can only use 1 pJ of energy. Another example is ambulatory healthcare monitoring, where vital signs of a patient are monitored over an extended period of time. Because these devices have to be mobile and small, energy is very limited. An added constraint is that the compute platform has to be programmable, as the field of ambulatory healthcare is still developing, and improved algorithms and new applications are developed at a fast rate.

2.1 Intrinsic Compute Efficiency

To support embedded applications like 4G communication, a computational operation has an energy budget in the sub-pico joule domain. However, current programmable devices do not have a high enough compute efficiency to meet this requirement. One of the most compute-efficient microprocessors, the ARM Cortex-M0 has a compute efficiency of 5.3 pJ/op at 40 nm low-power technology [2], not taking into account any memory accesses. A read operation on a 64 kilobyte memory consumes approximately 25 pJ of energy in 40 nm technology (based on the memory datasheets). Since the Cortex-M0 has no instruction caches, each cycle an instruction is loaded from the instruction memory. Therefore, the actual energy efficiency is closer to 30 pJ per operation, when memory accesses for data are

© The Author(s), under exclusive license to Springer Nature Switzerland AG 2022
M. Wijtvliet et al., *Blocks, Towards Energy-efficient, Coarse-grained Reconfigurable Architectures*, https://doi.org/10.1007/978-3-030-79774-4_2

ignored. The intrinsic compute efficiency (ICE) of 45 nm technology is 1 pJ/op [3], as can be observed in Fig. 2.1.

There is thus a large gap between the Intrinsic computational efficiency (ICE) and the achieved efficiency of at least a factor 30. This is because branching, instruction fetching and decoding, memory operations, and address calculations all contribute to overhead. However, to support compute intensive embedded applications, processors more powerful than the Cortex-M0 are needed, which increases the gap up to several orders of magnitude [3]. This can be attributed to the memory hierarchy as well as performance enhancing extensions that are added to the processor, such as branch prediction and dynamic out-of-order scheduling. In order to meet the demands of modern embedded applications, this gap has to be closed. Other works find a gap of (500×) between high performance general purpose processors and Application Specific Integrated Circuits (ASICs), which come very close to the ICE. There are many sources of inefficiency for general purpose processors that contribute to this gap. Hameed et al. identified various of these sources [5]. They extend a Tensilica processor [6] with complex instructions and configurable hardware accelerator to support a specific target application. This brings the compute efficiency for the application within 3× of the ICE, but at the expense of generality. The resulting architecture is highly specialized and is achieved in a step-wise approach. First, they apply data parallelism wherever possible to reduce instruction overhead, providing an order of magnitude improvement in energy efficiency. Second, the function units are optimized by adjusting the bit-width and instruction optimization, achieving another gain of a factor two. To provide further gains, they specialize the architecture completely towards their

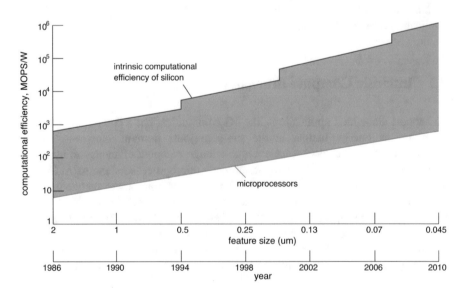

Fig. 2.1 Intrinsic compute efficiency for various technology nodes [4]

benchmark set, eventually achieving a programmable architecture that is a factor three less energy efficient than a dedicated ASIC. Although this architecture is very energy efficient and still somewhat programmable, it is very specialized towards specific applications. In other words, it lacks the flexibility that a CGRA can provide.

Based on their optimizations, it can be concluded that largest sources of overhead are:

1. Dynamic control, e.g., fetching and decoding instructions.
2. Data transport, e.g., moving data between compute, memory, caches, and register files.
3. Mismatch between application and architecture operand bit-width, e.g., 8-bit add on 32-bit adder.

The first source of overhead can be attributed to sequential execution. A large amount of energy is used because the processor fetches and decodes a new instruction every cycle. This can be mitigated by using spatial layout (where execution is spread in space, rather than in time). By increasing the number of issue slots, it is possible to achieve a single instruction steady state, such that no new instructions need to be fetched for an extended period of time. Figure 2.2 illustrates this by an example. The loop body contains operations A, B, C and control flow computation 'CF'. The loop in the figure can be transformed from the sequential version (left) to the spatial version (right) by software-pipelining. The single-cycle loop body in Fig. 2.2 does not require any other instructions to be fetched and decoded. It can be observed that a general purpose processor with only one issue slot can never support single-cycle loops due to the control flow. This technique is already used in very long instruction word (VLIW) processors [7], but is only applicable if the number of issue slots and their compute capabilities match the loop. ASICs and FPGAs implement an extreme form of spatial layout. By completely spatially mapping the application, the need for instruction fetching and decoding is eliminated altogether.

The second source of inefficiency, data transport, is reduced substantially by adapting the data-path to the application in such a way that the Register file (RF) and memory system are bypassed as much as possible like in explicit data-paths [8]. The memory system and the RF are two of the main energy users in a processor. Thus, by keeping data in the data-path, the overall energy usage can be reduced

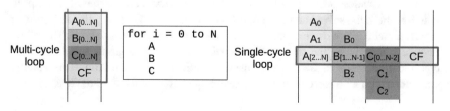

Fig. 2.2 Program execution example for multi- and single-cycle loops

significantly. If this is not possible, then care should be taken to keep the data in (multiple) small memories close to the function units.

The second source of inefficiency can be addressed by adapting the micro-architecture to the application. Applications have varying types of parallelism: Bit-level parallelism (BLP), Instruction-level parallelism (ILP), and DLP. BLP is exploited by multi-bit function units, such as a 32-bit adder. ILP is exploited with multiple issue slots, such as VLIW processors. Finally, DLP is exploited by SIMD architectures. Different applications expose different types and amounts of parallelism. When the micro-architecture is tuned to the application, such as in an ASIC or FPGA, the mix of different types of parallelism can be exploited in an optimal manner.

Multi-granular function units in an architecture can reduce the third source of inefficiency. These function units allow the required bit-width for an operation to be tuned by either splitting a wider unit into multiple narrower units, or by combining multiple narrower units into wider function units.

Micro-architecture adaptation is the key to achieve a higher compute efficiency. FPGAs and ASICs do this, but at an unacceptable price. For ASICs the data-path is adapted for one set of applications, so it loses generality. FPGAs are configured at gate level, which requires many memory cells to store the hardware configuration (bitfile). These cells leak current resulting in high static power consumption [9, 10]. Furthermore, the dynamic power is also high [11], due to the large configurable interconnect. Additionally efficiently compiling for FPGAs is hard due to its very fine granularity. Although there are High-Level Synthesis tools that reduce the programming effort, they cannot always provide high quality results [12] because of this. Summarizing, to achieve high compute efficiency, overhead of adaptability should be reduced, while still supporting:

1. Single instruction steady state, e.g., single-cycle loops.
2. Data transport reduction, e.g., explicit bypassing, local register files, and small memories close to function units.
3. Application tailored exploitation of parallelism, e.g., VLIW with matching SIMD vector lanes.
4. Programmability.

Part of these requirements are already investigated with existing CGRAs but there are still many open questions. Section 2.3 extensively investigates CGRAs in the past decades. Section 2.4 classifies their architectural properties. This helps to gain insight in the solved and open questions in this field. However, the definition of when an architecture can be considered a CGRA is not very well defined. Therefore, Sect. 2.2 introduces the definition that will be used throughout this book. The classification made in Sect. 2.4 is used to validate this definition. This chapter concludes with an overview of open research questions in the field of CGRAs.

2.2 Definition of CGRAs

There exist many informal and conflicting definitions of CGRAs. One work describes CGRAs as architectures that can only process loop bodies [13]. Another describes CGRAs as architectures that perform operations at a multi-bit level inside reconfigurable arrays [14]. However, the first definition excludes CGRAs that can execute full applications, and the second includes FPGAs with DSP blocks. A more generic definition that is used states that reconfigurable architectures use hardware flexibility in order to adapt the data-path at run-time to the application. They essentially provide a spatially programmable architecture [15, 16]. This requires that, before computation can be performed, some sort of configuration has to be loaded into the device. This configuration is typically static for some time, e.g. the duration of an application for an FPGA bitstream. However, virtually every computer architecture has a reconfigurable data-path to some extent. A typical general purpose processor uses cycle-based instructions to reconfigure the data-path to perform the desired operation, and even ASICs might have configuration registers that influence the design's operation.

Formally defining the CGRA class of architectures proves to be complicated, because almost all architectures are reconfigurable at some level. To get a better understanding, we observe that a computer architecture can be abstracted to the functional block shown in Fig. 2.3, which takes an input I, a configuration C, and uses these combined with an internal state S to produce an output O. The important part is that this abstraction can be made at various levels. If the data-path of a general purpose CPU is considered, then I is an input data stream, C is an instruction stream, and O is the data stream produced by executing the instructions on a cycle-to-cycle basis. In this case the Central processing unit (CPU) has reconfigured every cycle or instruction. However, if the whole CPU is considered, then I would be a dataset, e.g. an image, and C is the application code. Here the CPU would be configured once per application. To the user the functionality is identical, the granularity is invisible.

It is interesting to note here that the classification of the input into operands/data I or configuration C is merely an artificial classification, which has its origin in the way computer architects think about their designs. Often the definition of what is configuration and what is data is unclear, if not interchangeable. A good example is a LUT in an FPGA. When programmed to perform a binary operation, the contents of the LUT can be considered a configuration. But when the LUT is used as a small memory, it would be reasonable to define those same contents of the LUT as part of the input data I.

Architecture classes reconfigure their hardware at different levels, and in different temporal and spatial granularities (see Fig. 2.4). The size of the hardware blocks that are reconfigured, i.e., the level at which the system is observed according

Fig. 2.3 Abstraction of a
computer architecture

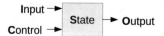

Fig. 2.4 Reconfiguration granularities in time and space

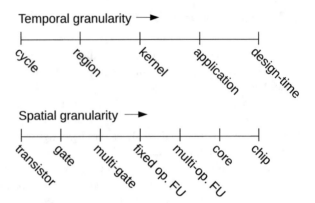

to the model described above, is the spatial domain of the reconfiguration. Spatial reconfiguration granularities used for defining CGRAs are:

- *Chip level* reconfiguration means that the entire chip design is changed per application. Effectively, each application has its own dedicated chip.
- *Core level* reconfiguration is used in multi-core processor systems where tasks can be mapped to one or more of these cores.
- *Multi-operation function unit level* is a typical reconfiguration level used in general purpose processor where an instruction reconfigures the ALU to perform a certain operation.
- *Fixed operation function unit level* is used in architectures where special acceleration units are configured to become part of a computation. An example thereof is a DSP block in FPGAs.
- *Multi-gate level* can be considered the level at which most FPGAs perform reconfiguration. An example of such a reconfigurable block is a look-up-table in an FPGA that allows implementation of multiple logic gates, such as a 3-input LUT, that can be used to implement logic functions.
- *Gate level* reconfiguration is performed for individual gates; these gates are not implemented using structures such as LUTs but consist of fixed implemented gates that can be connected by a reconfigurable network.
- *Transistor level* reconfiguration is quite uncommon because of the large overheads, but it is possible. An example thereof is the use of silicon nanowire transistors to construct configurable gates [17].

Besides granularity in the spatial domain, architectures can be reconfigured in the temporal domain. Granularity in the temporal domain specifies how long a configuration C is kept static before it is updated. The levels of temporal granularity considered for defining CGRAs are:

- *Design-time* reconfigurability means that configuration of the architecture is entirely static and cannot be changed after the architecture has been manufactured. This is typically the case for dedicated ASICs.

- *Application level* provides some level of reconfigurability. Devices providing application level reconfigurability are usually configured just before an application needs to be executed (e.g. at power up) and will run the same application for a very long time. This is typical for FPGAs.
- *Kernel level* reconfiguration allows an architecture to be reconfigured for sections of an application, for example a series of loop-nests required to perform a higher level function. Partial reconfiguration can be an example of kernel level reconfiguration.
- *Region level* reconfiguration operates at loop-nest level. Typically a small section of the application with a very high compute intensity. For each loop-nest, the architecture is reconfigured to provide good performance.
- *Cycle level* reconfiguration is performed by most processors that execute instructions. For each clock-cycle, the data-path is configured to perform the operation described by the instruction.

A CPU for example reconfigures the operation of an ALU (spatial granularity) at cycle level (temporal granularity), while an FPGA reconfigures LUTs (spatial) at application level (temporal). It is even possible to reconfigure at the transistor level [18] in the spatial domain, but in this work only gate level reconfiguration and coarser will be considered.

Considering the granularity of reconfiguration in both the temporal and spatial domain, it seems apt to define coarse-grained reconfigurable architectures as those architectures that reconfigure at a 'coarse' level. Based on the CGRAs discussed in Sect. 2.3 of this book, we propose to define an architecture to be classified as a CGRA if it possesses the following properties:

1. A spatial reconfiguration granularity at fixed functional unit level or above.
2. A temporal reconfiguration granularity at region/loop-nest level or above.

Since there can be many temporal and spatial levels of reconfiguration within the same system, as shown in Fig. 2.3, the focus should be on the dominant level when considering the properties above. Where the dominant level is the level at which (re)configuration has the greatest influence on the functionality of the system. For example, a CGRA can perform cycle-based instructions, but the width of these instructions is often several orders of magnitude smaller than static configurations at kernel or application granularity. Therefore, the kernel or application configuration is dominant. Figure 2.5 shows the architectural landscape for the dominant reconfiguration levels in both spatial and temporal dimensions. A third access is present describing the resource type that the two-dimensional space of temporal and spatial granularity describes. The resource type considered can be the data memory hierarchy, an interconnect network (if present), or the compute fabric. The figure shows where various computer architecture classes are located within the two-dimensional space with respect to their compute fabric. Dedicated ASICs are temporally configured at design time and spatially at the chip level, their operation does not change after the design has been manufactured and the whole chip is a dedicated layout. CPUs on the other hand typically reconfigure at the cycle

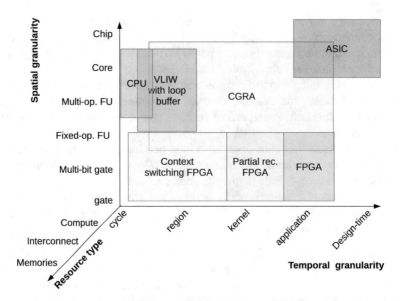

Fig. 2.5 Spatial-temporal architecture landscape. Horizontal and vertical axes indicate temporal and spatial granularity for reconfiguration. A third dimension allows specification of temporal and spatial granularity for various (sub)systems in an architecture, the figure shows classifications for the compute resources

level, each instruction their hardware performs another operation, this is done on the level of a processor core or ALU level. Processors that are extended with loop buffers may be reconfigured once per processing region, if a loop-nest fits entirely in the loop buffer then reconfiguration (at the processor level) is only required at region level. FPGAs provide reconfiguration at finer spatial granularity, typically from fixed operation function units (such as DSP blocks) to gate level reconfiguration. Temporal reconfiguration is performed over almost the entire temporal granularity range. There are FPGAs that provide context switching at region or even cycle level, those that provide partial reconfiguration at kernel level, and more static FPGAs that only reconfigure for an application.

Reconfiguration can be either performed statically or dynamically and at multiple levels of spatial granularity within the same architecture. The choices made at various spatial levels determine the distribution of configuration bits over the temporal domain. An FPGA has only coarse temporal granularity while a CPU has mostly fine temporal granularity. It is important to note that the CGRA principles can be applied to different subsystems. In many architectures for example, a compute, interconnect, and memory subsystem can be identified. For each of these subsystems, the coarseness of reconfiguration can be chosen differently, which sketches the vast design space of CGRAs (Fig. 2.5). Typically CGRAs differ from other architectures by allowing spatial mapping of applications over the available resources. To enable this, the interconnect subsystem has to be reconfigurable, causing many CGRAs to focus their reconfiguration options on the network. This is presumable also why Hartenstein decided that the interconnect is an important feature of CGRAs in his classification, already proposed more than a decade ago [14], a classification also adopted by a more recent survey [19].

2.3 A History of CGRAs

In the past, many CGRAs have been proposed and/or designed. This section summarizes a large selection of architectures that are either described by the authors of the original paper or referenced by other work as coarse-grained reconfigurable architectures. These architectures target various application domains, including digital signal processing, computer vision, encryption, and benchmarks such as SPECInt and Parboil. For each investigated architecture a brief summary will be provided. Based on this investigation, Sect. 2.4 classifies these architectures on specific architectural properties and analyses the efficacy of the proposed CGRA definition. The letters between square brackets (e.g. [A]) will be used to refer to the architectures in the remainder of this chapter. The figures shown in this section are based on the original papers; these papers are referenced in the text describing each architecture.

Xputer (1991) [A]
The XPuter architecture [20] consists of a Reconfigurable array (RA) of homogeneous units that are connected via a 2D mesh. The architecture distinguishes itself from most other CGRAs by the control flow. Instead of being controlled by a program counter based system, as shown in Fig. 2.6. It is controlled via a data sequencer and control tags associated to the data stream; unlike a dataflow architecture the data access pattern in memory is explicitly controlled by the sequencer. C-compiler and place-and-route tools for this architecture exist.

PADDI (1992) [B]
PADDI is built from several homogeneous FUs that are connected to a crossbar switch [21], as shown in Fig. 2.7. Each of these FUs contains an ALU for processing, and an instruction memory for control. The instruction memory is a 53-bit wide control word with eight entries. These entries can be different for each of the FUs

Fig. 2.6 Architectural structure of the Xputer architecture [20] showing the data sequencer and data memory. The 2D mesh network is located within the 'rALU' subsystem

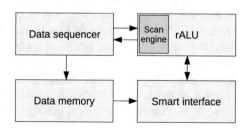

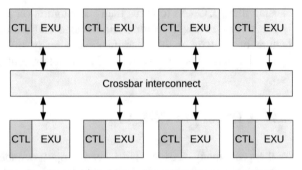

Fig. 2.7 The PADDI architectural structure [21], showing the crossbar switch and the function units connected to it. As shown, each function unit has its own instruction decoder (CTL)

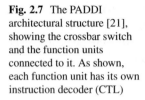

Fig. 2.8 PADDI-2 [22] has a modified architectural structure with respect to its predecessor. The network now contains two hierarchical levels

and act as one VLIW instruction. The active instruction is selected by a 3-bit index signal. The crossbar network is statically scheduled on a per cycle basis. The PADDI tool-flow provides a 'C-like assembly' language.

PADDI-2 (1993) [C]

PADDI-2 is the evolution of PADDI and improves on its predecessors short-comings [22]. In particular the centralized control of PADDI leads to instruction bandwidth problems and limits scalability. To overcome this, PADDI-2 (shown in Fig. 2.8) follows a data-flow model of execution, with local control in each processing element. Instead of FUs, nano-processors are constructed which each have their own instruction store and 3-bit program counter. The program counter allows eight contexts to be selected. Execution of an instruction on a nano-processor has no side-effects for other units, which enhances scalability. Furthermore, the crossbar of the original PADDI architecture is replaced by a 2-level hierarchical network, implemented by a semi-crossbar with switchable buses.

Fig. 2.9 The KressArray
architecture [23] contains a
grid of Xputer-like compute
elements that are connected
over a local neighbour
network. Similar to Xputer a
sequencer, the rDPA address
generation unit, is used to
perform data sequencing

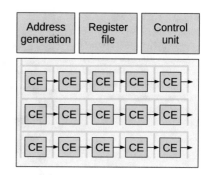

rDPA/KressArray (1995) [D]

The rDPA architecture [23], later renamed to KressArray, is strongly related to
the XPuter architecture. They use the same functional unit design but differ
in architectural organization. rDPA consists of an RA with local, neighbour to
neighbour, connections, as well as a global bus that connects all units to each other,
the outside world, and a global controller. The global controller performs address
generation and controls configuration loading. This requires full spatial layout as
the configuration is static per application. Data is provided as a stream under control
of the address generation unit, as shown in Fig. 2.9. The rDPA tool-flow includes
a C compiler and a custom hardware description language and compiler. Place and
route is performed via simulated annealing.

Colt (1996) [E]

Colt is an array of homogeneous 16-bit FUs connected through a toroidal mesh
network [24], as shown in Fig. 2.10. A 'smart (pruned) crossbar' connects this mesh
to the data ports of the chip. Each FU can perform integer operations and contains
a barrel shifter. FUs can be combined to support wider integer or floating point
operations. The Colt architecture is statically scheduled, and FUs include delays to
synchronize data with respect between FUs. The Colt has three modes of operation
for conditional execution, depending on the branch characteristics a specific mode
is selected. The idea behind this is to keep the control overhead to a minimum
for applications with simple control flow, but still provide support for applications
that require run-time reconfiguration coupled with complex looping structures and
conditional execution. Special units (DP) provide access to external memories as
well as any peripherals required for interfacing the system. A compiler is not
provided, but the Von Neumann execution model combined with programmable
delays to align different compute pipelines makes a compiler feasible.

MATRIX (1996) [F]

The MATRIX architecture [25] consists of a RA of homogeneous units, each with a
memory, ALU, control logic, and several network ports. The network contains three
levels, each in 2D layout, as shown in Fig. 2.11. Each network layer progressively
increases the distance spanned per connection from neighbourhood to global. The

Fig. 2.10 The Colt
architecture [24] consists of a
smart crossbar that connects
function units in a toroidal
structure. Special interfaces
(DP) provide data access to
external memories or
peripherals

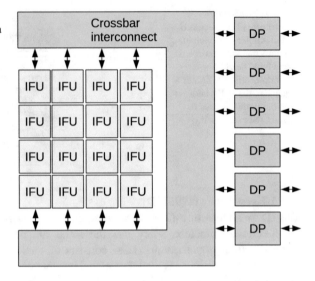

Fig. 2.11 The special layout
of the various interconnect
networks of Matrix [25] can
be observed in this figure

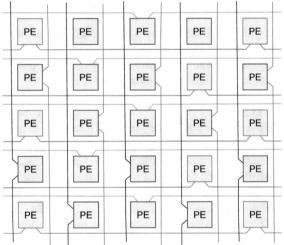

network is dynamically routed, but can be fixed by the configuration words. No
compilation flow for this architecture is described.

RaPID (1996) [G]

RaPID is a linear array of heterogeneous Processing elements (PEs) [26]. RaPID
mostly resembles a systolic array, but with a reconfigurable interconnect and
operation of the PEs (Fig. 2.12). This leads to a highly configurable architecture, but
also many control bits. RaPID solves this by defining 'hard' control, which is fixed
for the duration of an application (large temporal granularity), and 'soft' control,
which can be updated every cycle (fine temporal granularity). The division of bits in

Fig. 2.12 The RaPID [26]
architecture with its linear
array of function units and
memory elements. Data input
and output is performed via
the 'stream manager' that can
directly source and sink data
from the interconnect via
FIFOs

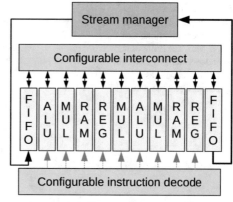

Fig. 2.13 Garp [28] provides
an array of 2-bit logic
elements that can be
combined horizontally to
form wider operations.
Multi-cycle operations can be
constructed vertically. These
operations can be used as ISA
extensions to a MIPS
processor

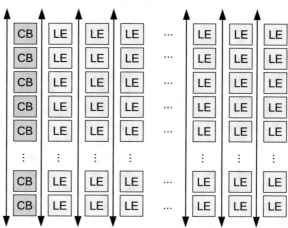

hard and soft differs per instance, but approximately 25% of the control is soft, and
75% hard. Furthermore, RaPID features hardware support for executing loop-nests
in an efficient manner. A compiler and programming language have been developed
for the RaPID architecture [27].

Garp (1997) [H]

The GARP architecture combines a MIPS processor with an FPGA-like accelerator [28]. The MIPS processor controls the configuration of the RA via extensions
to the MIPS instruction set. The RA consists of 2-bit logic elements that can form
operations column-wise and can communicate row-wise via a wire network, see
Fig. 2.13. By doing so the MIPS can offload more complex operations to the
RA, which provides Instruction set architecture (ISA) extensions to the processor.
Examples of such operations are dedicated operations for encryption and image
processing. Compilation for this architecture is performed via the GCC MIPS
compiler, the instruction set extensions are called from assembly sections.

Fig. 2.14 The RAW
architecture [29] consists of a
two-dimensional mesh of
RAW tiles. Each tile contains
an independent lightweight
processor as well as a per
cycle configurable network
switch

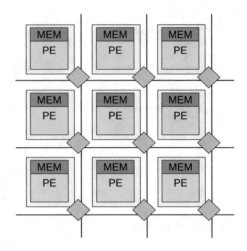

Fig. 2.14 The RAW
architecture [29] consists of a
two-dimensional mesh of
RAW tiles. Each tile contains
an independent lightweight
processor as well as a per
cycle configurable network
switch

RAW (1997) [I]

The RAW architecture [29] is a 2D array of RISC-style pipelined FUs, each with
their own instruction and data memory (shown in Fig. 2.14). The RAW architecture
therefore supports the MIMD execution model. These FUs can run independently
from each other and can communicate via a mesh network. Each FU has a memory
that controls the destination addresses used by the network interface, hence the
network access patterns can be statically scheduled. Each RAW tile contains
three memories. An instruction memory contains the instructions executed by the
ALU in each tile. A special memory (SMEM) is used to control the switch-box
configuration. The configuration is tied to the program counter and can therefore
update on a per cycle basis. Finally, there is a data memory present in each tile that
operates as a scratch-pad memory for data storage. The FUs have byte-granularity
and can be combined to form larger data types. A compiler based on a high-level
language is available, and place and route is TIERS based [30].

PipeRench (1998) [J]

The PipeRench architecture consists of several rows (stripes) of PEs [31]. PEs
in a row communicate through a neighbourhood network, and between rows a
crossbar type of network handles communication. PipeRench exploits 'pipeline
stage parallelism'. The stages of a pipelined loop are mapped to the available PEs.
If there are more stages than PEs, the PEs are run-time reconfigured to execute
different stages. Figure 2.15 shows the stages, called stripes, and their processing
elements. A compiler is available which takes a custom language (DIL) as input.
A greedy place and route algorithm was developed that runs much quicker than
available commercial place-and-route tools, but still yields acceptable results due to
the coarse grain nature of the architecture.

Fig. 2.15 PipeRench [31] contains multiple pipelined levels for processing elements with reconfigurable networks in between. Within a level processing elements can communicate to neighbouring processing elements

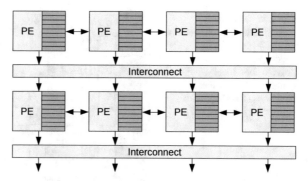

Fig. 2.16 REMARC [32] provides a grid of 16-bit function units that can be connected over a neighbourhood network as well as some longer distance connections

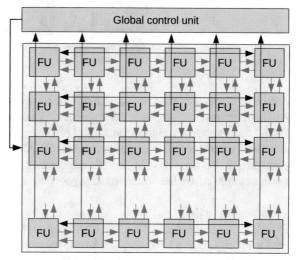

REMARC (1998) [K]

The REMARC reconfigurable architecture serves as a co-processor to a MIPS host and communicates via a bus to the host processor [32]. The RA consists of a 64-FU grid of 16-bit FUs that are connected via neighbour to neighbour connections, as shown in Fig. 2.16. The RA is controlled by a global control unit that manages data and generates a global program counter. Each FU has an instruction memory indexed by the program counter, within each row or column the FUs can function in SIMD mode. For compilation GCC is used. The application includes a binary file containing the RA configuration. The configuration is generated based on addresses in an assembly language.

CHESS (1999) [L]

The appropriately named CHESS architecture defines a grid of 4-bit ALUs and switch-boxes in a chequerboard pattern [33], as shown in Fig. 2.17. This layout supports strong local connectivity between ALUs. 4-bit ALUs are small compared to those typically used in CGRAs, but the ALUs of CHESS can be combined to support wider operations. An interesting feature of the switch-boxes is that they

Fig. 2.17 The Chess
architecture [33] consists of a
grid of 4-bit ALUs in a
chequerboard pattern to
provide good local
connectivity. Distributed
memories (RAM) provide
storage for intermediate
results

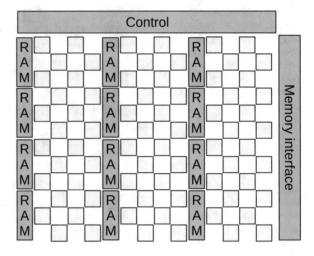

can be configured to work as local memories. This, combined with the ability of
ALUs to get instructions from the data network, allows the construction of 'mini-
CPUs' that execute a program from local memory, controlled by a global program
counter. A compiler does not exist, but is feasible. Complicating factors for compiler
construction are dealing with the aforementioned 'mini-CPUs', and routing data
when multiple ALUs are combined to support wide operations.

Pleiades (2000) [M]
Pleiades [34] defines an architecture that is composed of a programmable microprocessor which acts as a host, and heterogeneous compute elements (satellites) which
are connected via a reconfigurable interconnect, as shown in Fig. 2.18. The satellites
follow a data-driven execution model, which has support for vector operations. As
a proof of concept, the Maia processor was designed according to the Pleiades
template. Maia is focused on voice processing and uses a hierarchical mesh network,
and also features an embedded FPGA. Automated code generation for Maia is not
trivial, but has been done to some extent [35].

Chameleon/Montium (2000) [N]
The Chameleon architecture is built on several VLIW-like processing units connected via a circuit switched network and can be interfaced via a AHB-NoC
bridge [36]. Each processor has VLIW-like control and layout. Connections between
the inputs and outputs of compute units and local memories can be routed via
a crossbar network, as shown in Fig. 2.19. Configurations are stored in a local
instruction memory and are statically scheduled at compile-time. These configurations effectively perform a form of instruction expansion; instructions issued to
the 'instruction decoding' unit are decoded based on a preloaded configuration.
Therefore, depending on the configuration of the decoder and the configuration

Fig. 2.18 The 'satellite processors' processors of Pleiades [34] provide dedicated arithmetic acceleration to an embedded processor. Communication is performed via a reconfigurable interconnect

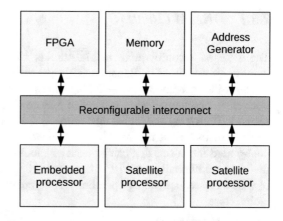

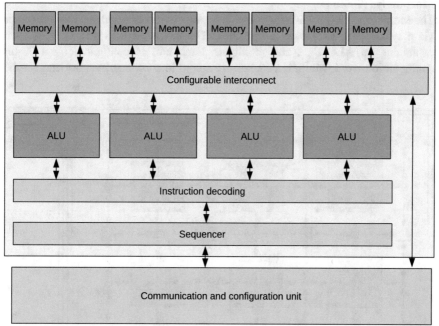

Fig. 2.19 The Montium [36] processor provides a configurable interconnect between processing elements and local memories. This network can be configured to form an application dependent data-path. An instruction sequencer and decoder provide VLIW-like control

of the ALU interconnect, instructions can result in different operations being performed (similar to a writeable controls store). Chameleon is designed by ReCore systems and has a C-compiler.

2.3.1 DReaM (2000) [O]

The DReaM reconfigurable architecture is highly specialized towards mobile wireless protocols [37]. Reconfigurable processing units (RPUs) are grouped in clusters of four, as shown in Fig. 2.20. Inside a cluster there is local interconnect, and inter-cluster communication is handled by a configurable SRAM-based switch-box network. Both local and global interconnects use a handshaking protocol. Each RPU consists of two 8-bit reconfigurable arithmetic processing units (RAPs), and two dual-ported memories that can be used as a look-up-table to aid multiplication and division. Additionally each RPU has a spreading data-path unit for fast execution of CDMA-based spreading operations. A compiler is not provided for the DReaM architecture.

MorphoSys (2000) [P]
The MorphoSys [38] RA is a reconfigurable processor connected to a host processor via a system bus, as shown in Fig. 2.21. The RA is an 8×8 grid of FUs, each containing an ALU, RF, and multiplier. Each unit is configured by a 32-bit configuration word. Multiple contexts are stored in a context memory and can be

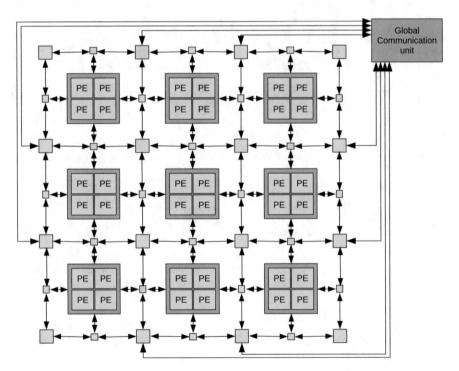

Fig. 2.20 The DReaM [37] architecture provides reconfigurable processing units grouped together in clusters. The figure shows a grid of 3x3 clusters. Clusters can communicate via a switch-box network

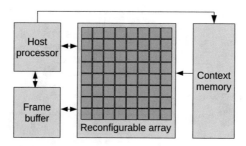

Fig. 2.21 MorpoSys [38] operates as an accelerator to a RISC processor. The accelerator consists of an 8×8 grid of function units that can interact with a frame buffer. A context memory provides multiple selectable configuration contexts for the array

Fig. 2.22 The Chimaera [39] provides bit-level reconfigurability via a reconfigurable array. Chimaera is intended to be integrated into the data-path of a host processor to provide instruction set extensions

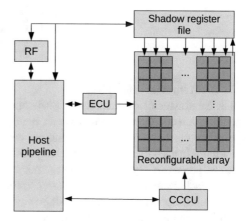

broadcast row- or column-wise to provide SIMD functionality. The network consists of four quadrants that are connected in columns and rows. Inside each quadrant, a more complex direct network is present. For the MorphoSys architecture, a C compiler is developed, but partitioning is performed manually.

Chimaera (2000) [Q]

Chimaera pushes the bounds on what can be considered a coarse grain reconfigurable architecture due to the bit-level granularity within the data-path [39]. The architecture defines an RA of logic cells that is integrated into the pipeline of a host processor, as shown in Fig. 2.22. The logic cells perform a reconfigurable operation at the bit level (like FPGAs), but inside the array they are organized in word-wide rows because the architecture primarily targets word-level operations. A shadow register file allows access of up to nine operands. This allows the definition of custom operations in the host processor. The configuration for a custom operation is stored in the memory of the host and can be cached locally for fast reconfiguration. Compilation can be done by selecting sequences of operations and mapping these to a single combined operation on the RA.

Fig. 2.23
Smartmemories [40] is a
reconfigurable architecture
with respect to its memory
hierarchy. Multiple small
memories can be connected
via a crossbar interconnect to
provide application specific
memory depths and widths

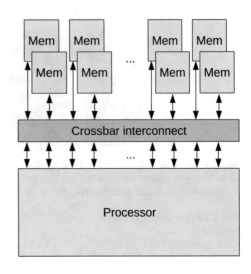

SmartMemories (2000) [R]

The SmartMemories architecture [40] consists of one or more processor tiles on a chip which communicate via a packet switched network, as shown in Fig. 2.23. Four tiles are clustered into a 'quad' and are connected via a faster local network. The memory in each tile can be connected in various hierarchies (via a crossbar network) to the processor. By doing so the memory hierarchy can be adapted to the application. Each memory has an address generator and can be controlled via instructions. Compiler support is not mentioned; however, this could be implemented based on a MIPS compiler with memory mapped interfaces to the crossbar network.

Imagine (2002) [S]

The Imagine stream processor contains eight VLIW clusters that operate in SIMD mode and are under control of a microcontroller [41], as shown in Fig. 2.24. Each cluster contains a configurable network that allows direct connection of ALU inputs and outputs, as well as a scratchpad memory and communication unit, much like the Chameleon [36]. All ALU clusters get their inputs from a stream memory. FUs start processing as soon as their input data is ready, thereby indicating a data-flow oriented architecture. Loop support is managed via a control processor. Streaming-C compiler support for this architecture has been developed.

ADRES (2003) [T]

The ADRES architecture [42] is a very tightly coupled VLIW and CGRA. The CGRA shares resources (including the register file and VLIW FUs) with the VLIW processor and can therefore run in either VLIW or CGRA mode, as shown in Fig. 2.25. In CGRA mode the RA can access the VLIW's register file, reducing transfer cost. The connections between the FUs in the RA are direct (no switch-boxes); however, ADRES is an architecture template and therefore these

Fig. 2.24 A stream register file provides data to the eight VLIW clusters present in the Imagine [41] stream processor. A control processor manages loops

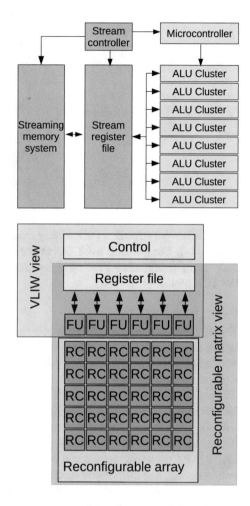

Fig. 2.25 The ADRES [42] architecture can operate in two 'views'. In one view ADRES operates as a standard VLIW processor, the other view allows operation of ADRES as a CGRA

connections can be modified at design time. ADRES features a C compiler for both VLIW and CGRA.

DART (2003) [U]

DART [43] defines clusters of 'reconfigurable data-paths' (DPRs), as shown in Fig. 2.26. A DPR consists of FUs connected via a reconfigurable, entirely connected, multibus network. Both the fully connected network, and the operation of the FUs, can be configured. This results in many control bits for one DPR (122 bits per DPR, excluding bits for the interconnect between DPRs). The goal of DART is to keep the configuration static for as long as possible, but when a change in functionality is required, a technique similar to the 'hard' and 'soft' configuration bits in RaPID [26] is employed. DART allows a full update of all configuration bits, but also partial updates. The partial updates do not affect the network, but only the operations of the FUs, and the operand sources. A compiler is not available, but

Fig. 2.26 Dart [43] contains multiple reconfigurable data-paths (DPR). Each DPR is managed by a controller that allows partial reconfiguration of the operations performed by function units inside each DPR

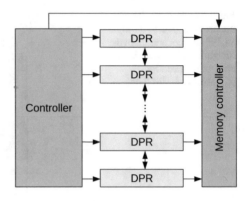

the multibus network in a DPR highly resembles a transport triggered architecture (TTA) [44]. Compilation for TTAs is certainly not trivial, but compilers do exist, such as the retargetable TTA compiler based on LLVM which is available through the TCE project [45].

PACT_XPP (2003) [V]
PACT_XPP defines two types of processing array elements (PAE), one for computation, and one with local RAM [46]. The PAEs are connected with a packet based network, and computation is event driven. Both the operation of the PAEs and the network are reconfigurable, which results in a large number of configuration bits. The control flow is handled in a decentralized fashion such that a configuration can be kept static as long as possible, Fig. 2.27 shows the central configuration controller (SCM) and the local configuration memories (CM) in each compute cluster. To support irregular computations that do require to update the configuration, PACT_XPP uses two techniques: configurations are cached locally to enable fast configuration, and *differential* configurations are supported. Differential configurations only update selected bits, optimizing the use of the local configuration cache. This is similar to the 'hot' and 'cold' bits in the DART architecture. The data-flow model of computation plus hardware protocols that guarantee that packets are never lost ensures that reconfiguration is transparent to the user, making compilation quite easy.

WaveScalar (2003) [W]
The WaveScalar architecture is a dataflow based reconfigurable architecture [47] that contains a pool of processing elements, as shown in Fig. 2.28. These processing elements are assigned instructions dynamically from a 'work-cache'. Scheduling instructions is dynamic and out of order, but takes dependencies into account according to a data-flow description of the application. The architecture contains four network layers, from local direct connections between processing elements, to routed connections between processing clusters. The compiler for WaveScalar is able to translate imperative programs (such as C) into a data-flow description; this description is also used as the target code. The out-of-order scheduling approach

Fig. 2.27 PACT_XPP [46] provides multiple clusters of processing elements. A central controller manages the configuration memories (CM) for each cluster

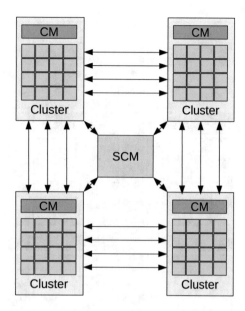

used in WaveScalar is good for performance as the available function units could be more efficiently utilized. However, the infrastructure needed to support out-of-order processing will lead to a lower energy efficiency.

TRIPS (2004) [X]

TRIPS is different from most CGRAs by its out of order (OoO) data-driven nature [48]. The architecture contains several OoO grid processors, as shown in Fig. 2.29. Each of these processors contains 16 issue slots that can communicate via a routed mesh network. Each processor dynamically schedules frames of operations onto the RA, and a frame cache keeps the most recently used frames for reuse. The TRIPS IIRC Compiler maps basic block data-flow graphs (DFG) to the CGRA and is based on the Imagine framework. Place and route is done within the frame-level and is performed at design time, but the processing resources for the frame itself are assigned dynamically at run-time.

Kim (2004) [Y]

Kim et al. propose a reconfigurable architecture [49], as shown in Fig. 2.30, that is based on MorphoSys [38], but with a notable exception. Where MorphoSys supports only the SIMD model, the proposed architecture also supports loop pipelining similar to PipeRench [31]. As a consequence the configuration of the reconfigurable compute elements (RC) can no longer be broadcast, but has to be stored locally, as each RC can perform a different operation. No compilation flow is available. However, compilation for this architecture would be about as difficult as it is to compile for MorphoSys, for which a compiler exists.

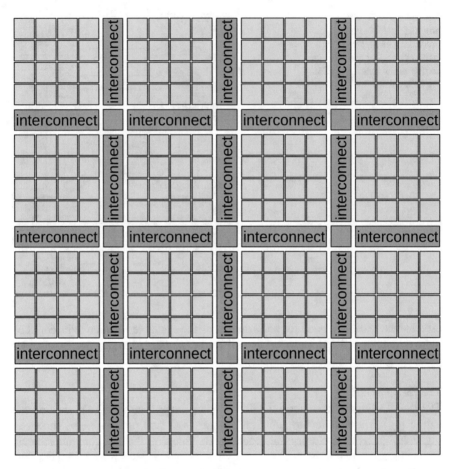

Fig. 2.28 The WaveScalar [47] architecture provides a large grid of data-flow driven processing elements. Each processing element contains input control to provide synchronization between the out-of-order processing elements. Also included are input and output buffers to decouple the execution from other PEs

STA (2004) [Z]

The STA architecture [50] consists of several clusters of compute elements. Each of these clusters operates in SIMD mode to provide vector processing capabilities. Within each cluster several processing elements with a fixed operation, such as addition, multiplication, and register files are present. Figure 2.31 shows one of the function units of STA; multiple of these operate in parallel within each cluster. The output of each processing element is connected to the input of all other processing elements. On a cycle basis a VLIW-like instruction activates one or more clusters (of SIMD units) and selects which of the inputs will be used for computation. By doing so, a spatial mapping of the algorithm can be created. The STA architecture is supported by a GCC based back-end that translates MatLab code into executables.

Fig. 2.29 The TRIPS [48] architecture contains multiple out-of-order grid processors. The figure shows two types of two of these processors, each with their own memory hierarchy containing several local memories and an interface to off-chip memory

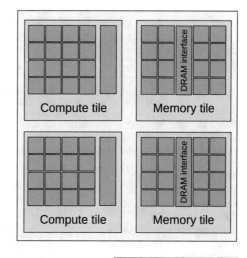

Fig. 2.30 The Kim [49] architecture operates as an accelerator to an ARM host processor. An AHB bus provides the accelerator access to the memory hierarchy

Fig. 2.31 The function units of STA [50] allow selection of operands from multiple TTA-style bus networks. In contrast to TTA, instructions are provided synchronously via a control signal interface

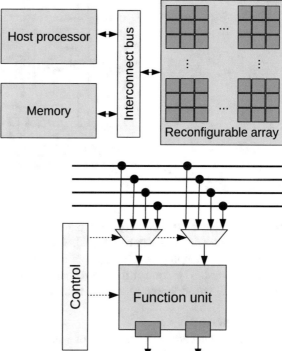

Galanis (2005) [AA]

Galanis et al. define a reconfigurable data-path [51], which is integrated in a SoC together with a host processor and embedded FPGA, as shown in Fig. 2.32. Compute nodes inside the reconfigurable data-path consist of an ALU and a multiplier, and they can take operands from other nodes or a register bank. The interconnect

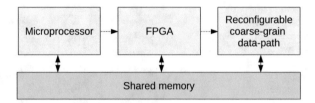

Fig. 2.32 The Galanis [51] architecture provides acceleration to a microprocessor. An FPGA and a CGRA interact with this microprocessor through a shared data memory. The FPGA can be used to provide bit-level acceleration as well as provide control signals to the CGRA

Fig. 2.33 One CLB of Astra [52] with its surrounding periphery. Astra operates on an 8-bit data-path. The control flow is managed via a 1-bit control flow network

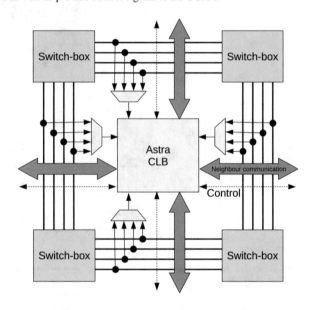

between the nodes is a full crossbar, but the authors note that if scalability is an issue, a fat-tree network can be adopted. The compute nodes and rich interconnect allow for easy mapping of sub-graphs of an application DFG, making compilation straightforward. The control for the reconfigurable data-path is generated by the embedded FPGA, providing sufficient flexibility to support complex control flow.

Astra (2006) [AB]

The Astra architecture features an RA of 8-bit FUs that can be combined to function as wider units [52]. These FUs contain multiple static configurations that can be changed at run-time. These contexts are selected via a separate control network that, although not described in the paper, could be controlled by a host processor or by internal logic. The network between FUs is a neighbour to neighbour network, and FU resources can be used to route over longer distances. A compilation flow is not described, but will be similar to place and route on an FPGA. Figure 2.33 shows an Astra function unit.

Fig. 2.34 An array of processing elements in the CRC [53] architecture. Each PE contains context memories that contain configurations, these configurations are selected by the state transition controller

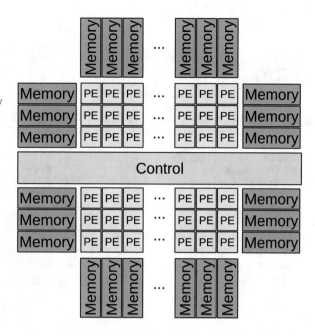

CRC (2007) [AC]

The configurable reconfigurable core (CRC) is an architecture template for 'processor-like' reconfigurable architectures [53]. In particular NEC's DRP[54] is an instance of this class of architectures. The architecture follows the typical CGRA template with an array of PEs with a reconfigurable interconnect (shown in Fig. 2.34) which is not further specified by the CRC architecture. One of the key features of the CRC architecture are context memories inside the PEs which enable fast switching of configurations, similar to those in PACT_XPP and Tabula [55]. This fast switching of contexts enables 'a third dimension for routing by redirecting communication through the time domain'. For compilation the Cyber [56] framework is used, and 'techniques for simultaneous scheduling, placement, and routing are developed on the basis of integer linear programming [57]'.

TCPA (2009) [AD]

TCPA [58] consists of an array of heterogeneous FUs, that resemble lightweight VLIW processors, connected via a neighbourhood network (Fig. 2.35). The heterogeneity of the FUs is a design-time decision. The interconnect between the FUs is statically configured, and forms direct connections between the FUs. Each FU has a (horizontal and vertical) mask that allows individual reconfiguration of FUs. In this way, SIMD type behaviour can also be implemented. Unlike conventional VLIW processors, the register files in these FUs are explicitly controlled. Compilation for the architecture is performed based on a description in PARO, a dependency description language, and in later work on the PAULA language.

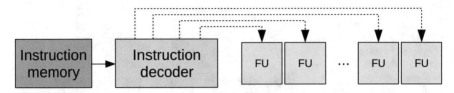

Fig. 2.35 Each processing element inside the TCPA [58] architecture contains multiple VLIW-style FUs. These FUs each have their own instruction memory and decoder

Fig. 2.36 The 4×4 array of PPA [59] contains PEs that each have their own local register file and can select multiple network sources as input operands. In addition, PPA allows selection of the central register file as one of its operands

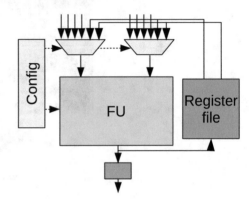

PPA (2009) [AE]

The polymorphic pipeline array (PPA) consists of clusters of four PEs called a core [59]. The four PEs inside a core share an instruction cache, loop buffer, and register file for predicate flags. Each PE has a register file, and an FU for arithmetic integer operations. Furthermore, one PE per core can perform a multiplication. Inside a core the FUs are connected through a mesh network, as shown in Fig. 2.36. For inter-core communication there exists three options, one of which is a neighbourhood network that links not the FUs, but the register files of adjacent cores. Another distinctive feature is the shared memory per x-column of PEs through a memory-bus, where x is configurable to be 1, 2, or 4. By having the amount of sharing configurable, the load latency can be kept low for applications that do not require sharing between cores. The compiler for PPA uses virtualized modulo scheduling to create schedules that can be virtualized on the available hardware resources.

CGRA Express (2009) [AF]

CGRA Express is an architectural extension [60] to the ADRES architecture. It adds the ability to perform latch-less chaining of function units to construct complex single-cycle operations. The units are chained together via a bypass network, Fig. 2.37 shows the internal structure of such a function unit. A modulo scheduling based compiler is extended to find common sub-graphs and combine operations with sufficient clock slack into a single cycle.

Fig. 2.37 The function units of CGRA express [60] extend the ADRES architecture with bypasses

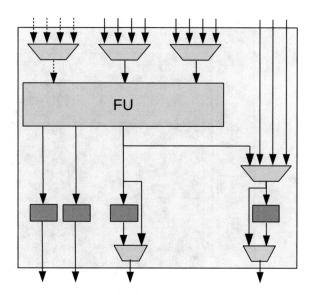

KAHRISMA (2010) [AG]

The KAHRISMA [61] architecture studies the level of reconfigurability that is required to support different types of applications. Based on their findings an architecture with mixed-granularity is introduced. The architecture has a grid of heterogeneous PEs consisting of both fine-grained and coarse-grained building blocks. Each coarse-grained PE had its own context memory and a local sequencer, implementing a form of lightweight processor. The fine-grained PEs can be used to implement special functions, similar to an FPGA. These PEs can therefore operate as small accelerators to the coarse-grained PEs. Binding of instructions to specific PEs is performed at run-time.

One of the innovations of KAHRISMA is the use of an extra level in the instruction fetch hierarchy that allows typical processor ISAs, such as the RISC ISA and a VLIW ISA, to be decoded and dispatched to KAHRISMA PEs. This allows KAHRISMA to have a flexible instruction set architecture. KAHRISMA provides a tool-flow that supports C and C++ programming languages as an input. Based on an intermediate representation and profiling results, sub-graphs are selected to be implemented as custom instructions. These instructions are then fed to a compilation back-end and further processed by binary utilities into machine code (Fig. 2.38).

EGRA (2011) [AH]

The EGRA architecture [62] is constructed as an array of ALU clusters, as shown in Fig. 2.39, called RACs. These RACs contain a 2D structure of ALUs that have a switch-box network between ALU rows. This amounts to a CGRA architecture with two network levels, a local network between ALU rows, and a neighbour network between clusters of these ALUs. The ALU clusters contain both local memories, a

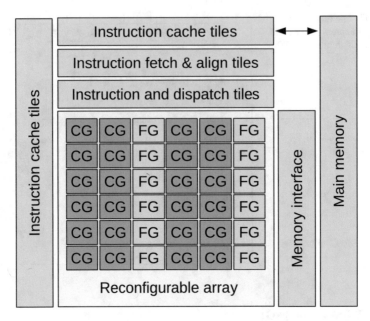

Fig. 2.38 KAHRISMA [61] contains a grid of both coarse-grained and fine-grained PEs

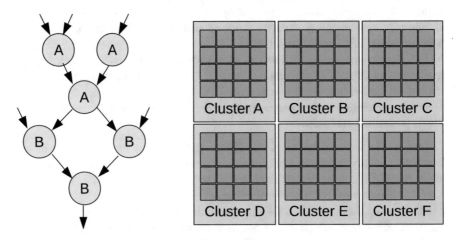

Fig. 2.39 The EGRA [62] provides a network of ALU clusters instead of providing only a single ALU per processing element. These clusters allow mapping of processing sub-graphs onto single PEs

larger memory, a multiplier, and ALUs. A cluster is therefore heterogeneous, while the cluster array is homogeneous. A compiler is not available, but the paper shows a manual method for application mapping that could be automated.

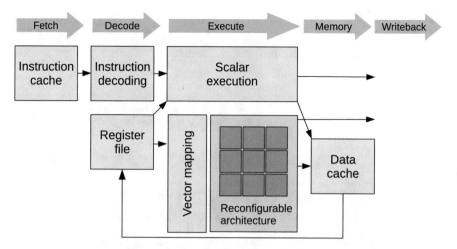

Fig. 2.40 DySER is integrated into the processing pipeline of an out-of-order processor. The figure shows a 3×3 array of processing elements that are designed to switch on a per cycle bases. This allows application specific ISA extension

DySER (2012) [AI]
The DySER architecture moves reconfigurable computing into the processor pipeline of an out-of-order processor [63]. With DySER an extra lane is added to the execution stage, as shown in Fig. 2.40. This issue slot contains a FIFO structure to construct vectors from sequentially arriving data elements, and then feeds these to an RA. The DySER RA does not support loops by itself, but loops can be performed via the host processor. The compiler for DySER is a modified Sparc compiler that can perform sub-graph matching to recognize DySER operations, and maps these to the circuit switched RA.

FPCA (2014) [AJ]
The FPCA architecture has a 2D grid of clusters [64], as shown in Fig. 2.41. Each cluster contains several compute elements, local memories, and a crossbar network. Each compute element performs same, fixed, operation (Similar to an FPGA DSP block). The actual operation is specified by providing constants. For example, a function unit with a multiplier followed by an addition can disable the addition by supplying the value zero as one of the operands. These constants and input values are routed between compute elements in the cluster by means of the crossbar network. The network between clusters is a neighbour to neighbour network. A scheduler performs dynamic resource allocation on the cluster array, and tries to replicate an application over the array until no more resources are available. An LLVM based compiler back-end is implemented which also performs placement. Routing is performed at run-time.

Fig. 2.41 FCPA [64]
provides acceleration via a
linear array of compute
elements (CE). These
function units can be
interconnected over a
'permutation data network' to
each other and to a memory
hierarchy

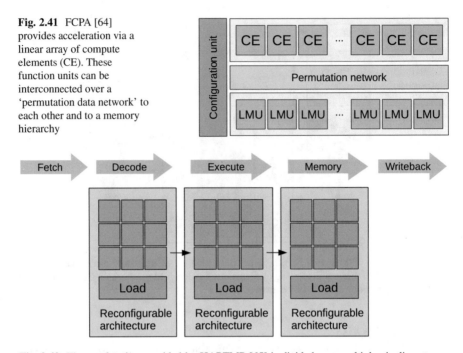

Fig. 2.42 The acceleration provided by HARTMP [65] is divided over multiple pipeline stages.
By doing so, the reconfigurable array can work in parallel to the execution scheme of a RISC
processor

HARTMP (2016) [AK]

HARTMP [65] is a reconfigurable matrix inside a RISC processor, as shown in
Fig. 2.42, allowing the ISA to be extended dynamically. Instructions are dynamically and transparently mapped to the RA. These RA configurations are stored in a
configuration cache for future use. When an instruction pattern matches a previously
executed RA configuration, it will get invoked instead of the instruction sequence.
When the RA is activated the pipeline of the RISC processor will be stalled
until processing on the RA is completed. This architecture has some similarities
to DySER [63], but adds dynamic configuration generation. Since configuration
generation is transparent, no compiler changes are required.

HyCUBE (2017) [AL]

HyCUBE [66] focuses on optimizing energy efficiency of CGRAs by optimizing
the interconnect between function units. The function units are connected in a two-
dimensional mesh topology similar to most CGRAs. However, instead of using
dedicated switch-boxes the routing elements are part of the function units and
under control of the instruction, as shown in Fig. 2.43. Therefore, the routing is not
static but can be changed at each cycle. This reduces the interconnect complexity
and therefore reduces consumed energy. However, the application must allow easy

Fig. 2.43 Each function unit within the HyCUBE [66] architecture can communicate with its neighbours. Communication is controlled per cycle

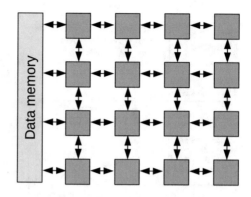

spatial layout otherwise passing values through the function units may become the bottleneck. The schedule for doing so is determined at design time by a compiler. Their interconnect provides a power reduction of 22% with respect to a standard (routed) network-on-chip. The processing elements within HyCUBE consist of an ALU with a cycle-based configuration memory.

X-CGRA (2019) [AM]
X-CGRA [67] incorporates approximate computing into CGRAs. Their function units allow configurable levels of approximation which is determined by the control word. A special field in the opcode, the 'OM' field, specifies the operating mode. Each operating mode specifies a specific quality output level. Figure 2.44 shows the architecture of X-CGRA. The operations and the required quality level provided by the function units are determined at design time by a custom scheduling and binding tool.

BioCare (2021) [AN]
The BioCare [68] architecture aims at energy efficiency for bio-signal processing. It does so by allowing configurable levels of approximation inside the processing elements. These processing elements are able to perform typical arithmetic operations, including addition, multiplication, and division. An optimization algorithm analyses properties of an application, with respect to energy and output quality, and determines the level of approximation for the PEs. Based on the executed kernel, the processing elements can be configured to SISD or SIMD mode. Figure 2.45 shows the architecture of BioCare.

2.4 CGRA Classification

The previous section summarizes a large selection of architectures that are either described by the authors of the original paper or referenced by other work as coarse-grained reconfigurable architectures. These architectures target various application domains, including digital signal processing, computer vision, encryption, and

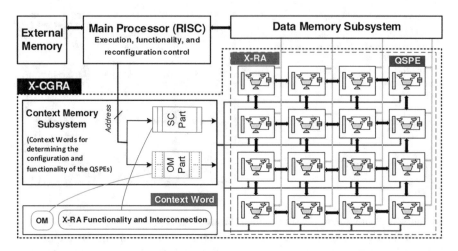

Fig. 2.44 An array of Quality-Scalable Processing elements as provided by the X-CGRA [67] architecture. A context memory subsystem controls the functionality and interconnect of the reconfigurable array as well as the 'operation mode' of each of the QSPEs

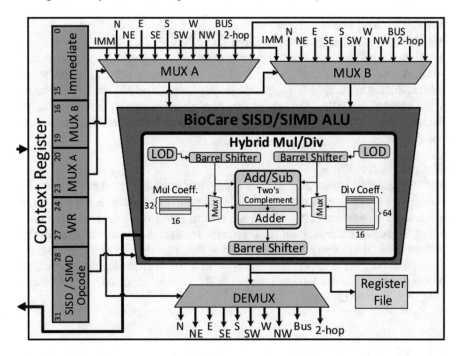

Fig. 2.45 BioCare SIMD CGRA [68] has PEs that can be configured for various levels of approximation. Depending on the required bit-width and precision it is possible to perform intra-PE data-level parallelism

benchmarks such as SPECInt and Parboil. For each investigated architecture a brief summary is provided. Based on this investigation, this section classifies these architectures on specific architectural properties and analyses the efficacy of the proposed CGRA definition. Each architecture is identified with one or two letters between square brackets (e.g. [A]) that will be used to refer to the architectures in the remainder of this chapter.

The overview shows that many CGRAs have been developed in the past decades. Each of these architectures adds their own new features and design choices. However, to properly understand what has been researched and where new research opportunities can be found the current CGRA landscape should be classified. Such a classification can be based on several architectural properties, which can be divided into four main categories: (1) the structure of the CGRA, (2) how it is controlled, (3) how it is integrated with a host processor (if any), and (4) the available tool support. These four categories are described in more detail below.

Structure

Each CGRA has different design choices for what is considered to be the *structure* of the architecture. The structure describes:

- Various network layers (e.g. topology and routing).
- FU data-path (e.g. FU granularity, FU operation types, register files and custom operation support).
- Memory hierarchy (e.g. scratchpads and caches).

Control

Scheduling (partial) configurations, both in time and space, can be performed either dynamic or static. For example, a compiler could decide the reconfiguration order (static schedule) or leave this up to the device (dynamic scheduling). Similarly, a place and route tool can statically decide which physical resource will perform an operation, or this can be mapped dynamically at run-time. This holds for both the configurations of the FUs (including register files and local memories) as well as for network configurations.

Integration

CGRAs can be tightly or loosely coupled to a host processor, or do not need a host processor at all and support stand-alone operation such as TRIPS [48]. Some CGRAs such as ADRES [42] share resources with the host processor, others only feature a communication interface to the host processor.

Tool Support

Any architecture's success depends heavily on the available tool support. For CGRAs this is often not only a compiler but also a place and route tool. However, comparing CGRAs just on their *available* tools is not a fair comparison. If tools are not available, an estimation will be made how feasible it is to implement the required tools, based on comparison with other architectures.

2.4.1 Classification Legend

The architectural properties are used to classify the reviewed architectures. In order
to conveniently classify multiple properties in a single table, a legend will be used.
This legend is described in Table 2.1 and shows the categories and properties that
are used to classify each of the architectures. If a certain property is not described
by the authors of a specific CGRA, this is marked with a '−' sign.

Table 2.1 Classification table, and legend for Table 2.2

	Property	Abbr.	Description
Structure	Network topology	M	Mesh network
		B	Bus network
		C	Crossbar
		D	Direct connection (e.g. between FUs)
	Granularity	P	Design-time parameter
	Multi-granular	Y/N	Yes/No
		S	Sub-word parallelism
	Fixed RF data-path	Y/N	Yes/No
	Memory hierarchy	M	Shared memory
		C	Cache
		S	Scratch-pad
Control	Scheduling	S	Static (compile-time)
		D	Dynamic (run-time)
	Reconfiguration	S	Static (compile-time)
		D	Dynamic (run-time)
	Network	S	Static (compile-time)
		D	Dynamic (run-time)
	Custom operations	Y/N	Yes/No
Integration	Integration	S	Stand-alone
		A	Accelerator
	Coupling	T	Tightly coupled
		L	Loosely coupled
	Resource sharing	Y/N	Yes/No
Tooling	Compiler	Y/N	Yes/No
		I	Imperative language
		D	Dataflow language
		C	Custom language
	Place and route & DSE	Y/N	Yes/No

2.4.2 Classification

With the classification criteria defined and a notation established, the classification can be made. The reviewed architectures are listed in order of introduction year in Table 2.2. The four classification categories are listed horizontally, resulting in vertical columns that can be compared. Within each category, the more detailed classification criteria are listed.

The reviewed architectures were either described by the authors of the original paper or referenced by other work as coarse-grained reconfigurable architectures. Section 2.2 presented a definition for CGRAs. Since such a definition should align with the general perception of a CGRA, the architectures from Sect. 2.3 are analysed and classified in Fig. 2.46. As can be observed, there is a strong correlation between our initial CGRA classification region in Fig. 2.5 and the evaluated architectures, indicating that these guidelines represent the general consensus.

There are two notable exceptions in the reviewed architectures with respect to the proposed CGRA definition, these are Garp [H] and Chimaera [Q]. Garp uses 2-bit logic elements which are configured in a method quite similar to FPGAs. The reconfigurable logic is used to create instruction set extensions to a MIPS processor. This behaviour is indeed quite different from the proposed definition in which a CGRA uses coarse-grained building blocks to construct a data-path in which there is also some dynamic control, such as instructions. A similar argument holds for the Chimaera architecture which has a bit-level granular reconfigurable data-path that is integrated into the data-path of a host processor.

2.5 Observations on Past CGRAs

Based on the review of the previously mentioned architectures, several observations can be made. These observations are described in this section and will use the same categorization as used in the classification section. Based on these observations, guidelines for future CGRA development are proposed.

Structure
Direct neighbour-to-neighbour and 2D-Mesh networks are the most popular choices. Both result in a regular grid that has good scaling properties. Crossbar networks are used on architectures with a lower number of FUs, or are used in FU clusters within a larger design. Although most CGRAs operate on a word-level granularity, there are some architectures that are multi-granular. The advantage of a multi-granular architecture is that the data-path of the architecture can more easily adapt to the application at a lower level. Some architectures allow custom operations, either by supporting some bit-level reconfigurable logic within the data-path, or by latch-less operation chaining to form complex single-cycle operations. The memory hierarchy is not very well defined in most architectures. They either

Table 2.2 Architecture classification, refer to Table 2.1 for a legend

General		Structure	Data-path			Memory	Control				Integration			Tool support			
Architecture	Year of publication	Network type(s)	Granularity	Multi-granular	Register files	Memory hierarchy	Operation scheduling	(Partial) reconfiguration	Network configuration	Custom/chained operations	Stand-alone/accelerator	Coupling	Resource sharing	Compiler	Place and route	Automated DSE tools
Xputer [A]	1991	2D-M	32-bit	N	Y	–	S	S	S	N	S	–	–	D	Y	Y
PADDI [B]	1992	C	16-bit	Y	Y	–	S	D	D	Y	S	–	–	A	Y	Y
PADDI-2 [C]	1993	2D-M,D	16-bit	N	–	–	D	S/D	D	N	A	L	N	D	Y	Y
rDPA [D]	1995	B,D	32-bit	N	N	–	S	S/D	S	N	S	–	–	I	Y	Y
Colt [E]	1996	3D-M,C	16-bit	Y	–	–	S	D	D	Y	S	–	–	N	N	N
MATRIX [F]	1996	D,D,D	8-bit	Y	N	–	S/D	S/D	S/D	Y	S	–	–	N	N	N
RaPID [G]	1996	1D-B	P	Y	–	S	S	S/D	S/D	Y	A	T	N	I	N	N
Garp [H]	1997	D	2-bit	Y	N	M	S	S	S	Y	A	L	Y	I/A	Y	N
RAW [I]	1997	2D-M	8-bit, Bit	Y	Y	S	S	S	D	Y	S	–	–	I	Y	N
PipeRench [J]	1998	C,D	P	Y	Y	–	S	D	D	Y	A	T	N	C	Y	N
REMARC [K]	1998	D	16-bit	N	Y	S	S	S	D	N	A	L	N	A	N	N
CHESS [L]	1999	2D-M	4-bit	Y	Y	S	S	S	S	Y	A	L	N	N	N	N
Pleiades [M]	2000	D,M,M	P	N	–	S	D	D	S	Y	A	L	N	I	Y	N
Chameleon [N]	2000	C	32-bit	N	N	S	S	D	S	N	A	L	N	I	Y	N
DReaM [O]	2000	D,D	8/16-bit	N	–	S	S	D	D	N	S	–	–	N	N	N
MorphoSys [P]	2000	D,D	16-bit	N	N	S	S	D	S	N	A	L	N	I	N	N
Chimaera [Q]	2000	D,B	4-bit	Y	Y	S	D	D	D	Y	A	T	N	I	Y	N
Smart Memories [R]	2000	B,D,C	64-bit	N	N	S	D	D	D	N	S	–	–	N	Y	Y
Imagine [S]	2002	C	32-bit	N	N	S	S	D	D	N	S	–	–	N	Y	N
ADRES [T]	2003	D	P	N	N	C	S	D	D	N	A	T	Y	I	Y	N
DART [U]	2003	B	16-bit	N	–	S	D	D	D	Y	S	–	–	N	N	N
PACT_XPP [V]	2003	M	P	N	–	S	D	D	D	N	S	–	–	I/C	N	N

Architecture	Year	M,B,D				C										
WaveScalar [W]	2003	M,B,D	-	-	-	C	D	D	D	N	S	-	-	I	Y	N
TRIPS [X]	2004	2D-M	32-bit	N	Y	C,C	D	D	D	N	S	-	-	I	Y	N
Kim [Y]	2004	D,D	16-bit	N	N	S	S	D	S	N	A	L	N	N	N	N
STA [Z]	2004	C,C	-	N	-	S	S	S	D	N	S	-	-	I	Y	N
Galanis [AA]	2005	C	16-bit	N	Y	C,M	S	D	D	N	A	L	Y	N	N	N
Astra [AB]	2006	2D-M, D	8-bit	Y	-	-	S	D	D	Y	S	-	-	N	Y	N
CRC [AC]	2007	-	P	N	Y	-	S	D	D	N	S	-	-	I	Y	N
TCPA [AD]	2009	D/M/C/B	32-bit	N	N	-	S	S	S/D	N	S/A	L	N	D	Y	Y
PPA [AE]	2009	2D-M	32-bit	N	Y	S,M	S	D	D	N	A	L	N	I	Y	N
CGRA express [AF]	2009	D,D	N	N	N	C	S	D	D	Y	A	T	Y	I	Y	Y
KAHRISMA [AG]	2010	-	32/1	Y	Y	C	D	D	D	Y	S/A	L	-	I	Y	N
EGRA [AH]	2011	D,D	-	N	-	M,S	S	D	D	N	A	L	N	N	N	N
DySER [AI]	2012	2D-D	-	N	-	S	S	D	S	N	A	T	Y	I	Y	N
FPCA [AJ]	2014	2D-D,C	32-bit	N	N	M,S	S	D	D	N	A	L	N	I	Y	N
HARTMP [AK]	2016	2D-M,M	-	N	-	N	D	D	D	N	A	T	Y	I	Y	N
HyCUBE [AL]	2017	2D-M	-	N	-	-	S	S	S	N	A	-	-	C	Y	Y
X-CGRA [AM]	2019	2D-M	16-bit	N	Y	M	S	S	S	N	A	L	Y	D	Y	Y
BioCare [AN]	2021	2D-M	8/16/32-bit	Y	Y	M	S	S	S	N	A	L	Y	N	N	Y

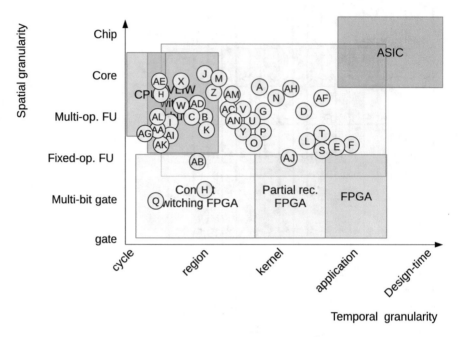

Fig. 2.46 Evaluated reconfigurable architectures and their spatial and temporal granularity

have some local scratch-pad memories, or a shared memory/cache, but few present more hierarchical structures.

Control

The architectures with a higher level of dynamic control are slightly more popular compared to architectures with more static control. This is especially true for dynamic (partial) reconfiguration, and the interconnect network. Therefore, most CGRA architectures aim towards high performance rather than very high energy efficiency. In order to reduce the required memory bandwidth for configuration loading and switching, some architectures feature configuration caches which reduce memory bandwidth requirements and allow fast configuration switches. Some architectures, such as RaPID [26] and Chameleon/Montium [36], allow a variable ratio between hard (more statically configured) and soft (cycle basis configuration). Doing so can be very beneficial for energy consumption.

Integration

CGRAs are both used as a stand-alone device, or as an accelerator. When used as an accelerator they tend to be loosely coupled to a host processor and communicate via a bus or memory interface. Some architectures, such as ADRES [42], share memories or compute resources. Other architectures are completely integrated into the host processor [63, 65].

Tool Support

Most architectures provide at least basic tool support such as an assembler, and place and route tools. A substantial number also provides a high-level language compiler. Often the compiler is C based, but domain specific language compilers (e.g. for data-flow descriptions) are also used. Some architectures feature an explicit data-path which makes compilation harder but can yield more energy-efficient architectures. Place and route is mostly performed at compile-time, although some architectures perform mapping and routing at run-time.

Most architectures do not provide any design space exploration tools, such as simulators and architecture models for power and performance estimation. Such tools can prove very useful for decisions between hard/soft configuration and interconnect variations.

Almost all architectures either have no compiler flow or have a dedicated way of describing an application. Only SRP [69], a variation on ADRES, uses a standardized programming model (OpenCL). This is an interesting field for investigation. High-level language support for CGRAs will improve portability between GPUs, FPGAs, and CGRAs, and might lead to a wider industrial acceptance of CGRAs.

2.5.1   Our Research Guidelines

Based on analysis of existing CGRAs five key topics for future research can be identified, which can be regarded as essential to unlock the full potential of CGRAs:

1. Only five of the 40 reviewed works present energy or power numbers, one of which compares power to signal processors (in 130 nm technology). This shows that CGRAs in the past have been focusing mainly at performance instead of energy efficiency.
2. CGRAs containing a dynamic configuration for both the compute and the interconnect are the most popular. These architectures can provide better application mappings. However, dynamic reconfiguration has a power disadvantage. The trade-off between application mapping and the level of reconfigurability is not well understood.
3. Most of the evaluated architectures focus on the compute and interconnect aspect of the CGRAs. Few architectures focus on the memory hierarchy despite that a significant part of the energy budget typically is spent on the memory.
4. Few architectures feature design space exploration tools and models of the architecture, which would allow design-time power and performance estimation of specific architecture configurations. These estimations can guide application developers to target low-energy or high-performance design goals.
5. Although many architectures feature some sort of tool-flow, very few are a commercial success. Often, the tool-flow is very architecture specific, thereby limiting industry acceptance. Programming models like OpenCL, supported by compilers, can help CGRAs compete with GPUs, and FPGAs.

In this book the focus will be on the first and second research topic, by (1) concentrating on energy efficiency and (2) providing a good balance between static and dynamic reconfiguration. Of course this cannot be done without paying some attention to programmability and the memory hierarchy as well. Furthermore, an architectural power and area model will be presented which can be a starting point for design space exploration tools.

2.6 Conclusions

An investigation into intrinsic compute efficiency shows that there can be a large difference between ASICs and general purpose processors. This difference can be as high as $500\times$. This gap is caused by various sources of inefficiencies that can be found in programmable architectures. Some of the main sources of inefficiency are: data movement, instruction fetching and decoding, and program control flow overhead. CGRAs can be used to reduce these inefficiencies.

However, before existing CGRAs can be investigated, it is first needed to be clear which architectures can be called a CGRAs. A definition to classify architectures as a CGRA is presented and defined as an architecture that has:

1. A spatial reconfiguration granularity at fixed functional unit level or above, and;
2. A temporal reconfiguration granularity at region/loop-nest level or above.

This definition is evaluated against a large selection of architectures that are either described by their authors as CGRAs, or by work referring to the original works as CGRAs. Based on this evaluation, it can be concluded that the definition accurately captures the general consensus of what a CGRA should be.

During this evaluation, several observations on existing CGRAs were made which can be translated into research guidelines. These research guidelines will be used to obtain a CGRA that is (1) more energy efficient and (2) provides a good balance between static and dynamic reconfiguration.

References

1. C.H. Van Berkel, Multi-core for mobile phones, in *Design, Automation Test in Europe Conference Exhibition, 2009. DATE '09* (2009), pp. 1260–1265. 10.1109/DATE.2009.5090858
2. ARM, *Cortex-M0 - Arm Developer*. https://developer.arm.com/products/processors/cortex-m/cortex-m0. Accessed 2019 June 14
3. A. Kumar et al., *Multimedia Multiprocessor Systems: Analysis, Design and Management*. Embedded Systems (Springer Netherlands, Dordrecht, 2010). ISBN: 978-94-007-0083-3. https://doi.org/10.1007/978-94-007-0083-3
4. A. Kumar, Analysis, design and management of multimedia multiprocessor systems. PhD thesis. Eindhoven University of Technology, 2009

5. R. Hameed et al., Understanding sources of inefficiency in generalpurpose chips. SIGARCH Comput. Archit. News **38**(3), pp. 37–47 (2010). ISSN:0163-5964. 10.1145/1816038.1815968
6. Cadence, *Tensilica Customizable Processor IP*. http://ip.cadence.com/ipportfolio/tensilica-ip. Accessed 2015 November 20
7. Y. Qian, S. Carr, P. Sweany, Loop fusion for clustered VLIW architectures. ACM SIGPLAN Notices **37**(7), 112–119 (2002)
8. L. Waeijen et al., A low-energy wide SIMD architecture with explicit datapath. J. Signal Process. Syst. **80**(1), 65–86 (2015). ISSN: 1939-8018. 10.1007/s11265-014-0950-8 http://dx.doi.org/10.1007/s11265-014-0950-8
9. F. Li et al., Architecture evaluation for power-efficient FPGAs. *Proceedings of the 2003 ACM/SIGDA Eleventh International Symposium on Field Programmable Gate Arrays*. FPGA '03 (ACM, New York, NY 2003), pp. 175–184. ISBN: 1-58113-651-X. 10.1145/611817.611844. http://doi.acm.org/10.1145/611817.611844
10. L. Deng, K. Sobti, C. Chakrabarti, Accurate models for estimating area and power of FPGA implementations. *IEEE International Conference on Acoustics, Speech and Signal Processing, 2008. ICASSP 2008*, March 2008, pp. 1417–1420. 10.1109/ICASSP.2008.4517885
11. A. Amara, F. Amiel, T. Ea, FPGA vs. ASIC for low power applications. Microelectron. J. **37**(8), 669–677 (2006) ISSN: 0026-2692. http://dx.doi.org/10.1016/j.mejo.2005.11.003. http://www.sciencedirect.com/science/article/pii/S0026269205003927
12. M. Wijtvliet, S. Fernando, H. Corporaal, SPINE: from C loop-nests to highly efficient accelerators using Algorithmic Species, in *25th International Conference on Field Programmable Logic and Applications (FPL), 2015*, September 2015, pp. 1–6. 10.1109/FPL.2015.7294015
13. B. Sutter, P. Raghavan, A. Lambrechts, Coarse-grained reconfigurable array architectures, in Handbook of Signal Processing Systems (Springer US, Boston, MA, 2010). ISBN: 978-1-4419-6345-1
14. R. Hartenstein, A decade of reconfigurable computing: a visionary retrospective, in *Proceedings of the Conference on Design, Automation and Test in Europe* (IEEE Press, Piscataway, NJ, 2001)
15. R. Tessier, K. Pocek, A. DeHon, Reconfigurable computing architectures, Proc. IEEE **103**(3) (2015). ISSN:0018-9219
16. P.E. Gaillardon, *Reconfigurable Logic: Architecture, Tools, and Applications*. Devices, Circuits, and Systems (CRC Press, Boca Raton, 2015). ISBN: 9781482262193
17. M. Raitza et al., Exploiting transistor-level reconfiguration to optimize combinational circuits, in *Design, Automation Test in Europe Conference Exhibition (DATE), 2017*, March 2017, pp. 338–343. https://doi.org/10.23919/DATE.2017.7927013
18. P. Groeneveld, P. Stravers, Ocean: the sea-of-gates design system. Delft University of Technology. Faculty of Electrical Engineering, 1993
19. V. Tehre, R. Kshirsagar, Survey on coarse grained reconfigurable architectures. Int. J. Comput. Appl. **48**(16), 1–7 (2012)
20. R.W. Hartenstein et al., A novel ASIC design approach based on a new machine paradigm. IEEE Journal of Solid-State Circuits **26**(7), 975–989 (1991)
21. D.C. Chen, J.M. Rabaey, A reconfigurable multiprocessor IC for rapid prototyping of algorithmic-specific high-speed DSP data paths. IEEE J. Solid-State Circ. **27**(12), 1895–1904 (1992). ISSN: 0018-9200
22. A.K.W. Yeung, J.M. Rabaey, A reconfigurable data-driven multiprocessor architecture for rapid prototyping of high throughput DSP algorithms, in *Proceeding of the Twenty-Sixth Hawaii International Conference on System Sciences, 1993*, vol. 1 (IEEE, New York, 1993)
23. R.W. Hartenstein, R. Kress, A datapath synthesis system for the reconfigurable datapath architecture, in *Design Automation Conference, 1995. Proceedings of the ASP-DAC'95/CHDL'95/VLSI'95, IFIP International Conference on Hardware Description Languages. IFIP International Conference on Very Large Scal* (IEEE, 1995)
24. R. Bittner, M. Musgrove, P. Athanas, Colt: an experiment in wormhole run-time reconfiguration, in *Photonics East Conference on High-Speed Computing, Digital Signal Processing, and Filtering Using FPGAs* (1996)

25. E. Mirsky, A. DeHon, MATRIX: a reconfigurable computing architecture with configurable instruction distribution and deployable resources, in *IEEE Symposium on FPGAs for Custom Computing Machines, 1996. Proceedings* (IEEE, New York, 1996)
26. D.C. Cronquist et al., Architecture design of reconfigurable pipelined datapaths, in *20th Anniversary Conference on Advanced Research in VLSI, 1999. Proceedings* (IEEE, New York, 1999)
27. D.C. Cronquist et al., Specifying and compiling applications for RaPiD, in *IEEE Symposium on FPGAs for Custom Computing Machines, 1998. Proceedings*, April 1998
28. J.R. Hauser, J. Wawrzynek, Garp: a MIPS processor with a reconfigurable coprocessor, in *The 5th Annual IEEE Symposium on Field-Programmable Custom Computing Machines*, April 1997
29. E. Waingold, Baring it all to software: Raw machines. Computer **30**(9), 86–93 (1997)
30. C. Selvidge et al., TIERS: Topology Independent Pipelined Routing and Scheduling for VirtualWire Compilation, in *Proceedings of the 1995 ACM Third International Symposium on Field-programmable Gate Arrays, FPGA '95* (ACM, Monterey, CA, 1995). ISBN: 0-89791-743-X
31. S.C. Goldstein et al., PipeRench: a reconfigurable architecture and compiler, in *Computer*, April 2000
32. T. Miyamori, K. Olukotun, REMARC: Reconfigurable multimedia array coprocessor. in *IEICE Transactions on Information and Systems*(1999)
33. A. Marshall et al., A reconfigurable arithmetic array for multimedia applications, in *Proceedings of the 1999 ACM/SIGDA Seventh International Symposium on Field Programmable Gate Arrays* (ACM, New York, 1999)
34. H. Zhang et al., A 1-V heterogeneous reconfigurable DSP IC for wireless baseband digital signal processing. IEEE J. Solid-State Circ. **35**(11), 1697–1704 (2000)
35. M. Wan et al., Design methodology of a low-energy reconfigurable single-chip DSP system". J. VLSI Signal Process. Syst. Signal Image Video Technol. **28**, 1–2 (2001)
36. G.J.M. Smit et al., The Chameleon architecture for streaming DSP applications. EURASIP J. Embedded Syst. **2007**, 078082 (2007). ISSN: 1687-3955
37. A. Alsolaim et al., Architecture and application of a dynamically reconfigurable hardware array for future mobile communication systems, in *2000 IEEE Symposium on Field-Programmable Custom Computing Machines* (IEEE, New York, 2000)
38. H. Singh et al., MorphoSys: an integrated reconfigurable system for dataparallel and computation-intensive applications. IEEE Trans. Comput. **49**(5), 465–481 (2000). ISSN: 0018-9340
39. Z.A. Ye et al., CHIMAERA: a high-performance architecture with a tightlycoupled reconfigurable functional unit, in *Proceedings of the 27th International Symposium on Computer Architecture, 2000*, June 2000
40. K. Mai et al., Smart Memories: a modular reconfigurable architecture, in *Proceedings of the International Symposium on Computer Architecture* (2000)
41. U. Kapasi et al., The imagine stream processor, in *Proceedings 2002 IEEE International Conference on Computer Design*, September 2002
42. B. Mei et al., ADRES: an architecture with tightly coupled VLIW processor and coarse-grained reconfigurable matrix, in *Field Programmable Logic and Application* (Springer, New York, 2003)
43. R. David et al., DART: a dynamically reconfigurable architecture dealing with future mobile telecommunications constraints, in *Résumé* (2003)
44. H. Corporaal, *Microprocessor Architectures: From VLIW to TTA* (Wiley, New York, 1997)
45. CPC group, TTA Based Co-Design Environment. http://tce.cs.tut.fi/. Accessed 2016 April 04
46. V. Baumgarte et al., PACT XPP - A self-reconfigurable data processing architecture. J. Supercomput. **26**(2), 167–184 (2003)
47. S. Swanson et al., The WaveScalar architecture. ACM Trans. Comput. Syst. **25**(2), 1–54 (2007). ISSN: 0734-2071

48. K. Sankaralingam et al., Trips: a polymorphous architecture for exploiting ILP, TLP, and DLP, in *ACM Transactions on Architecture and Code Optimization* (2004)
49. Y. Kim et al., Design and evaluation of a coarse-grained reconfigurable architecture, in *Proceedings of the ISOCC04* (2004)
50. G. Cichon et al., Synchronous transfer architecture (STA), in *Computer Systems: Architectures, Modeling, and Simulation* (Springer, New York, 2004)
51. M.D. Galanis et al., A reconfigurable coarse-grain data-path for accelerating computational intensive kernels. J. Circ. Syst. Comput. **14**(04), 877–893 (2005)
52. A. Danilin, M. Bennebroek, S. Sawitzki, Astra: an advanced space-time reconfigurable architecture, in *International Conference on Field Programmable Logic and Applications, 2006. FPL '06*, August 2006
53. T. Oppold et al., CRC-concepts and evaluation of processor-like reconfigurable architectures (CRC-Konzepte und Bewertung prozessorartig rekonfigurierbarer Architekturen), in *IT-Information Technology (vormals IT+ TI)*, vol. 49(3) (2007)
54. H. Amano et al., Techniques for virtual hardware on a dynamically reconfigurable processor an approach to tough cases, in *Field Programmable Logic and Application: 14th International Conference, FPL 2004, Proceedings*, Leuven, August 30–September 1, 2004 (Springer, Berlin, Heidelberg, 2004). ISBN: 978-3-540-30117-2
55. S. Teig, *Going beyond the FPGA with Spacetime*. http://www.fpl2012.org/Presentations/Keynote_Steve_Teig.pdf. Accessed 2016 April 12
56. K. Wakabayashi, T. Okamoto, C-based SoC design flow and EDA tools: an ASIC and system vendor perspective, in *IEEE Transactions on Computer-Aided Design of Integrated Circuits and Systems* (December 2000). ISSN: 0278-0070
57. J.A. Brenner et al., Optimal simultaneous scheduling, binding and routing for processor-like reconfigurable architectures, in *International Conference on Field Programmable Logic and Applications, FPL* (2006)
58. H. Dutta et al., A Holistic approach for tightly coupled reconfigurable parallel processors. Microprocess. Microsyst **33**(1), 53–62 (2009). ISSN: 0141-9331
59. H. Park, Y. Park, S. Mahlke, Polymorphic pipeline array: a flexible multicore accelerator with virtualized execution for mobile multimedia applications, in *Proceedings of the IEEE/ACM International Symposium on Microarchitecture. MICRO 2009*. ISBN: 978-1-60558-798-1
60. Y. Park, H. Park, S. Mahlke, CGRA express: accelerating execution using dynamic operation fusion, in *Proceedings of the 2009 International Conference on Compilers, Architecture, and Synthesis for Embedded Systems*. CASES '09 (ACM, Grenoble, 2009). ISBN: 978-1-60558-626-7
61. R. Koenig et al. KAHRISMA: a novel hypermorphic reconfigurable-instruction-set multi-grained-array architecture, in *2010 Design, Automation Test in Europe Conference Exhibition (DATE 2010)*, March 2010, pp. 819–824. https://doi.org/10.1109/DATE.2010.5456939
62. G. Ansaloni, P. Bonzini, L. Pozzi, EGRA: a coarse grained reconfigurable architectural template. IEEE Trans. Very Large Scale Integration (VLSI) Syst. **19**(6), 1062–1074 (2011). ISSN: 1063-8210
63. V. Govindaraju et al., Dyser: unifying functionality and parallelism specialization for energy-efficient computing, in *Micro* (2012)
64. J. Cong et al., A fully pipelined and dynamically composable architecture of CGRA, in *2014 IEEE 22nd Annual International Symposium on Field-Programmable Custom Computing Machines (FCCM)* (IEEE, New York, 2014)
65. J.D. Souza, L.C.M.B. Rutzig, A.C.S. Beck, A reconfigurable heterogeneous multicore with a homogeneous ISA, in *DATE* (2016)
66. M. Karunaratne et al., HyCUBE: a CGRA with reconfigurable single-cycle multi-hop interconnect, in *2017 54th ACM/EDAC/IEEE Design Automation Conference (DAC)*, June 2017, pp. 1–6. https://doi.org/10.1145/3061639.3062262
67. O. Akbari et al., X-CGRA: an energy-efficient approximate coarse-grained reconfigurable architecture, in *IEEE Transactions on Computer-Aided Design of Integrated Circuits and Systems* (2019), pp. 1–1. ISSN: 1937-4151. https://doi.org/10.1109/TCAD.2019.2937738

68. Z. Ebrahimi, A. Kumar, BioCare: an energy-efficient CGRA for bio- signal processing at the edge, in *ISCAS* 2021 (2021)
69. H.S. Kim et al., Design evaluation of OpenCL compiler framework for Coarse-Grained Reconfigurable Arrays, in *Field-Programmable Technology (FPT)*, December 2012. https://doi.org/10.1109/FPT.2012.6412155

Chapter 3
Concept of the Blocks Architecture

The existing CGRAs are often implemented either as coarse-grained FPGAs where execution of an algorithm is performed as a static systolic array like structure, such as Xputer [1] and ADRES [2] in CGRA mode or as a network of lightweight processors, like WaveScalar [3] and TCPA [4].

Systolic arrays provide a grid of function units that are configured statically to perform a specific operation for the duration of the application. Figure 3.1 shows an example of a systolic array. Although systolic arrays lead to energy-efficient implementations with good performance, these CGRAs are rather static and do not support operations that influence the control flow of an application, such as branching. The flexibility of these architectures, therefore, is low. Due to the static nature of the implementation there is no time-multiplexing of function units that may lead to a larger area than CGRAs that support cycle-based operations.

The second method uses a configurable network with multiple processing elements that operate independently or in lock-step. This method can be considered as very flexible but the power draw is much higher. This leads to a lower energy efficiency and is caused by the cycle-based operation of the processing elements. Execution control of these processing elements can either be performed by a global 'context' that is switched out when the schedule requires it (Fig. 3.2a) or local decoding per processing element (Fig. 3.2b). The execution schedule of local decoding can be controlled by a global program counter or by local execution control. The latter brings these architectures closer to a networked many-core than a CGRA, see Fig. 3.2c.

Blocks has a different structure compared to previous CGRAs that allows run-time construction of processors that match the type of parallelism found in applications very accurately, see Fig. 3.2d. By doing so, the energy overhead of reconfiguration can be reduced. At the same time, the flexibility of Blocks is kept. The structure that allows for this will be further described in Sect. 3.1. Section 3.2 describes the structure of the Blocks interconnect network. Section 3.3 introduces the memory hierarchy. Section 3.4 details how configuration of Blocks

© The Author(s), under exclusive license to Springer Nature Switzerland AG 2022
M. Wijtvliet et al., *Blocks, Towards Energy-efficient, Coarse-grained Reconfigurable Architectures*, https://doi.org/10.1007/978-3-030-79774-4_3

Fig. 3.1 Example of a
systolic array. The
connections between the
processing elements are either
fixed, or configurable at
application start-up, but do
not change during execution.
The same holds for the
operations the processing
elements perform. Data is
performed in a streaming
(pipelined) fashion and
'flows' through the array
during processing

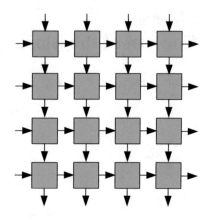

is performed. In Sect. 3.5 the function units available in the Blocks framework are
introduced. Section 3.6 details how various processor structures can be constructed.
Sections 3.7, 3.8, and 3.9 describe three concepts (multi-processor instantiation,
multi-granular function units, and approximate computing) that can be applied to
the Blocks framework. Section 3.10 concludes this chapter.

3.1   Separation of Data and Control

The main goal of Blocks is to increase energy efficiency while keeping flexibility.
Since energy efficiency is defined as performance divided by power, it can be
increased in two ways: by speeding up the computation and by reducing power
consumption. Systolic arrays perform well on both of these aspects. Spatial mapping
of the computation provides an efficient and pipelined path to compute results,
resulting in high performance. The static nature of the interconnect and operations
provides good power efficiency, which combined with high performance leads to a
low-energy architecture. However, this only works when there are enough compute
resources to spatially map the application to the hardware. Additionally, it is usually
not possible to map applications with a complex, or data dependent, control flow
due to the lack of flexibility at run-time.

Traditional CGRAs, with control schemes such as those shown in Fig. 3.2a–
c, have higher energy consumption due to their cycle-based reconfigurations.
Reducing the overhead of the control and sequencing will reduce the overall energy
consumption of the architecture. Where application specific processors reduce
control overhead by supporting DLP, this is not common in CGRAs. If data
parallelism is supported, then this is usually accomplished by special vector units in
the data-path. Since the required vector width strongly depends on the application,
it is hard to determine what vector width to choose at design time if the goal is

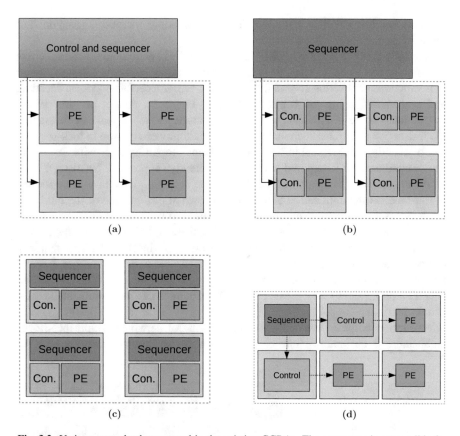

Fig. 3.2 Various control schemes used in the existing CGRAs. The sequencer is responsible for controlling the program flow, usually via a program counter or context number. Control (Con.) is responsible for loading the configuration or instructions into the processing elements (PE). For example, control can be an instruction decoder loading and decoding cycle-based instructions. (**a**) Global context switching. All processing elements are controlled simultaneously as one very wide instruction. This type of CGRA can be considered as a VLIW. (**b**) Local decoding with global sequencer. Each processing element has a local instruction memory and decoder. A global sequencer controls which instructions are fetched and decoded. Instructions, therefore, are performed in lock-step, resulting in a VLIW processor. (**c**) Local decoding and sequencing. The processing elements can run completely independently from each other. Effectively, every PE is a lightweight processor. This type of CGRA is very close to a many-core processor and provides an MIMD execution model. (**d**) Blocks' structure and the connections between sequencing, decoding, and function units are reconfigurable (indicated with dotted lines) and instantiated over a reconfigurable network. This allows the decoding structure to match the application. Multiple sequencers and decoders can be present, allowing multiple independent processors and support for multiple parallelism types

to keep the architecture as generic as possible. It is of course possible to switch off unused lanes in a vector unit but this creates area overhead that could have been used for other computations. Combining vector units to increase vector width means that there are multiple identical instruction fetch and decoding operations going on, thus diminishing the energy gains from DLP.

Virtually all CGRAs provide ILP in the form of VLIW style instructions. These instructions can be entire contexts (which can even be static, like in a systolic array) or instructions controlling several function units. CGRAs performing local decoding and sequencing, using the control scheme in Fig. 3.2c, do not perform VLIW instructions but are still able to perform instruction-level parallelism by constructing the instruction schedules such that the 'lightweight processors' run in lock-step.

To provide closer matching of the algorithm properties (e.g. with respect to vector operations) it is required not only to efficiently support ILP but also DLP. To efficiently and flexibly support DLP, SIMD-like configurations have to be formed at run-time, as the application is still an unknown at design time of the hardware in many cases. To do so, instruction decoders have to be dynamically connected to one or more function units. After configuration these function units are then under control of a single instruction decoder and operate in SIMD mode. This requires, besides a reconfigurable data network, a reconfigurable network to route decoded instructions to the function units, thereby creating a separation between control and data. Figure 3.3 shows how SIMD and VLIW structures can be implemented on the Blocks fabric.

Separation between the data-path and the control networks within the Blocks architecture is the main feature of Blocks that allows improvements on both the

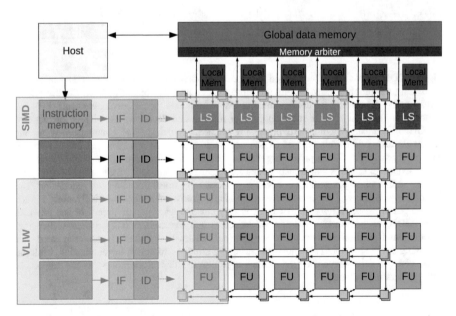

Fig. 3.3 Overview of the Blocks architecture. Showing how SIMD and VLIW structures can be implemented on the reconfigurable fabric. A function unit can be an ALU, load–store unit, a register file, a multiplier, or an immediate unit. The type of each function unit is chosen at design time

performance and the power aspects. The specialized data-paths in the constructed processors both reduce the number of cycles it takes to execute an application and reduce power by avoiding memory accesses due to increased bypass opportunities.

The structure of Blocks, as shown in Fig. 3.3, requires that every instruction decoder can control every type of function unit. Alternatively, it would have been possible to implement instruction decoders that are specialized per function unit. However, doing so reduces the flexibility of the architecture and increases complexity of the networks. This is because the total number of required instruction decoders will be higher to provide the same configuration opportunities. For example, two generic instruction decoders can control an ALU and a multiplier, or in another configuration control an ALU and an Load-store unit (LSU). To do so with specialized instruction decoders would require three decoders. Although specialized instruction decoders are simpler, the complexity of their connection to the network does not change. For this reason, Blocks uses generic instruction decoders. Every instruction decoder is able to control all types of function units.

The instruction decoders that Blocks uses are partially configured by the bit-stream, and hence per application or kernel. The configuration in the bit-stream sets which type of function units will be controlled by the instruction decoder (ID). For example: the ID can be set up to control an ALU or multiplier, but not at the same time. This is because the decoded instructions are function unit specific, and the bit-stream effectively configures which part of the instruction decoder is active. Per instruction decoder three bits are used to configure the type of function unit under control. These configuration bits effectively specify the highest bits of the instruction and select how the lower instruction bits will be decoded, as shown in Fig. 3.4. The three statically configured bits allow for eight types of function units to be controlled. If needed, it is easy to extend this to a larger number as the width of this field is a design parameter. Since the instruction decoders are configured to control a certain type of function unit, the opcodes in the input instructions can be reused. For example, the binary encoding for an 'add' instruction on the ALU can be the same as a 'multiply' instruction on the multiplier unit. This reduces the required width of the instruction stream, reducing the width of the instruction memories and, therefore, energy. Although there are some exceptions, most instructions have the format shown in Fig. 3.4. Not all fields are always required. The 'pass' instruction for example requires only one source and a destination to be specified.

To control program flow there is a need for a module that sequences the instructions on a processor configured on Blocks. In most CGRAs this is performed by a sequencer unit that is connected to all, or a subset of, function units on the

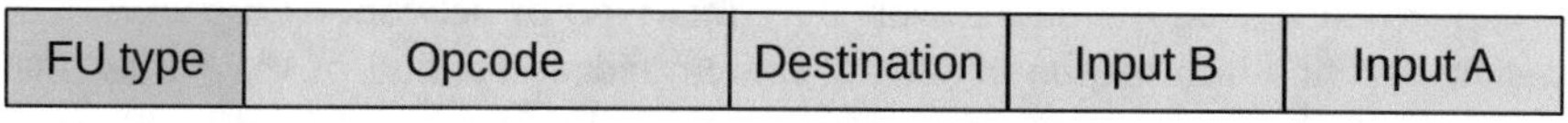

FU type	Opcode	Destination	Input B	Input A

Fig. 3.4 The most commonly used instruction format in Blocks. The green fields are dynamic fields and can be updated every cycle. The blue 'FU type' field is statically configured as part of the Blocks configuration

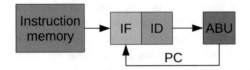

Fig. 3.5 Connection of an ABU to an instruction fetch (IF) and instruction decode (ID) unit. This is the minimal processor structure required to provide control flow for the application

reconfigurable fabric. This connection is typically fixed, thereby determining which function units will operate in lock-step. By not determining this at design time and allowing run-time configuration instead, construction of VLIW processors can be performed without design time assumptions on the number of issue slots. In fact, when multiple sequencers are available, it is even possible to construct multi-processor systems on the same fabric. The building block that controls sequencing of instruction decoders in Blocks architectures is called an accumulate and branch unit. This unit is available as a function unit on the compute fabric, like any other function unit. The Accumulate and branch unit (ABU) is used to manage the control flow by executing branching instructions. Branches can be conditional or unconditional. Conditions can be generated by any function unit on the fabric, for example by an ALU. Based on the supplied instructions and conditions the ABU then computes the next program counter value. This program counter determines the sequencing of the instruction decoders and is routed from the ABU to one or more Instruction fetcher and instruction decoders (IFIDs). Figure 3.5 shows the minimal processing configuration required to provide control flow for an application. Connecting the ABU to other function units allows for more complex control flow patterns. Since these IFIDs all receive the same program counter, they will operate in lock-step.

3.2   Network Structure

Most existing CGRAs include a configurable data network that can route data from one processing element to another, bypassing register files or memories. These networks can either be statically or dynamically configured. Statically configured means that they are configured once per application or kernel and keep this configuration for the duration thereof, similar to the networks used in FPGAs. A dynamically configured network can change during program execution, and an example thereof are routing networks. Instead of providing a direct path between a source and a sink, the data travels over this type of network as a packet. The packet contains information about the source and destination of the packet that is processed by routers that provide the connection between network segments. Although dynamic networks are more flexible, packets can move from any source to any destination, and they provide a higher overhead in terms of power, area, and latency (cycles).

Since the goal of Blocks is to reduce energy, a statically configured data network is used. The network configuration is part of the bit-stream that is used to configure the Blocks platform at run-time. Typically, the network is configured just before the start of an application, or a loop-nest within an application. Although the network is static during execution, dynamic CGRA behaviour is achieved by instructions issued to the function units. These instructions allow for dynamic input port selection.

In contrast to existing CGRAs, Blocks uses a separate network to connect the control signals from instruction decoders to function units. These control signals are decoded instructions and have, in most cases, a different width as the data. The control network that Blocks uses is, like the data network, a circuit switched network that is typically configured once per application or kernel.

Circuit switched networks typically use switch-boxes to route signals over a network. Switch-boxes can be compared to how a (wired) phone network operates (although more and more phone services are routed over Internet nowadays). In these networks, the phone number represents a configuration for the various switches on the line to make a connection between two selected devices. This connection then becomes a direct connection between these devices. Switch-boxes in Blocks work in a very similar manner; a bit-file provides a configuration for each of the switch-boxes on the network. This configuration contains specifications like: connect the left port of the switch-box to the bottom port of the switch-box, as shown in Fig. 3.6 with a solid green line. This figure shows a generalization of a switch-box as currently used in Blocks. The dots on the output ports are multiplexers that are configured to select one of the available inputs. The switch-boxes in Fig. 3.6 allow selection of all inputs that are not on the same side as the output. For example, it is not possible to use the left input as a source for the left output. The inputs and outputs on each side of a switch-box are referred to as channels.

Fig. 3.6 Generalization of a switch-box, as used in both the data and the control network. The outputs each have a multiplexer that selects from the available inputs (indicated with a dot). The solid green line represents a selected connection

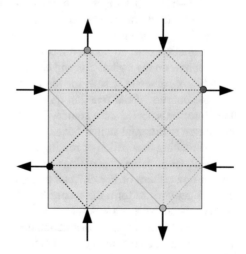

Although both control and data could, in theory, be routed over the same network, there are a few advantages to split one larger network into two separate networks.

First, the data-width of the control signals is a fixed (relatively small) width, while the data-width is a design parameter. It may, therefore, be wider or narrower than the control signals. In both cases, the maximum of both widths has to be taken. This causes wider than necessary switch-boxes that increase overhead. Second, the switch-boxes that perform data routing would require more ports to provide the same number of available data channels if control is also routed over the same switch-box. The number of routing options scales quadratic with the number of channels. For example, a switch-box with two input channels and two output channels on each side provides 48 routing options (six outputs times eight input options, as it does not make much sense to connect outputs to inputs on the same side of the switch-box). For a switch-box with three channels this already provides 108 routing options. Instead, providing two separate switch-boxes with two input channels (one for control and one for data) provides 96 routing options, as well as a lower propagation time of signals through the switch-box due to a less complex multiplexer structure.

The number of connections (channels) on each side of the switch-box is a design parameter. It is also possible for switch-boxes to have no inputs or outputs on one or more sides; this is generally the case for switch-boxes on the outside of the Blocks fabric. The difference of the Blocks switch-boxes compared to those found in FPGAs is that they switch a set of wires (a bus) instead of single wires. This reduces the amount of configuration bits that are required significantly. The Blocks tool-flow currently supports mesh-style networks but can very easily be extended to other network topologies. The width of the data network is a design parameter; the currently supported widths are 8-, 16-, 32-, and 64-bit wide data networks. This allows the designer to construct Blocks architectures with the appropriate data-width.

Switch-boxes that are close to function units can have inputs or outputs to these function units. In the Blocks framework this is typically represented by a diagonal connection, as shown in Fig. 3.7. The upper switch-box (grey) serves as an input to the function unit. To do so, it has an extra output. This output functions like any other output on the switch-box and can select any of the inputs available. The lower switch-box serves as a sink for the outputs of the function unit. On this switch-box the outputs of the function unit are available as a switch-box input and can, therefore, be selected as a source. The inputs and outputs of a function unit are connected to different switch-boxes. This allows to connect function units vertically while reducing occupation of routing resources.

Aside from some connections to function units and instruction decoders, the switch-box pattern of Fig. 3.7 can be replicated for both the data network and the control network to form the reconfigurable fabric of Blocks. The control network does not need to act as a sink for function units, while the data network does not need to sync any values from the instruction decoders. For both networks the interconnect properties such as the number of channels and the topology are a design parameter. Connections to and from function units are automatically generated by the Blocks tool-flow. An example fabric is shown in Fig. 3.8. The control network is shown in red. For this network it can be observed that instruction decoders (marked with

Fig. 3.7 Connection of function units to the Blocks reconfigurable network. The outputs of function units can be selected in the switch-box configuration, just like any other input on the switch-box. Inputs to a function unit are considered an output on a switch-box, allowing it to select any of the inputs as a source

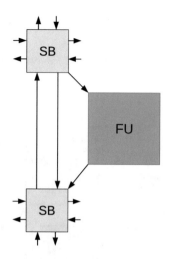

IFID) are only sources on this network, whereas function units are only sinks. The data network shown in blue has no such limitations. The data network is also used to connect the ABU to the instruction decoders in order to supply the program counter (more on this in Sect. 3.6.1).

3.3 Memory Hierarchy

Applications need data to process. In the many CGRA architectures reviewed in Chap. 2 only few report on the used memory hierarchy. However, the memory hierarchy can have a significant effect on the efficiency of the compute fabric of a CGRA. If the hierarchy is not able to supply data fast enough, the computation has to be stalled (paused). Stalls can never entirely be avoided in realistic memory hierarchies. This is especially the case in applications where memory accesses have poor spatial locality. This means that data required to perform a step in the computation cannot be grouped into a limited number of memory accesses. In applications with good memory locality it is often possible to group multiple accesses into one larger access to, ideally, one memory word. This allows multiple values to be loaded with only a single memory access; this grouping is called coalescing. Coalescing is especially important for accesses to external memories where the energy cost per access is high and the bandwidth that can be provided is comparatively low to on-chip memories.

Blocks provides a multi-level memory hierarchy. The highest memory level is called the global memory and is a memory that can be either a large on-chip memory or an external memory interfaced to the Blocks architecture. A Blocks architecture instance contains one or more load–store units. Each of these units can load data from the global memory. This means that there can be multiple load and store

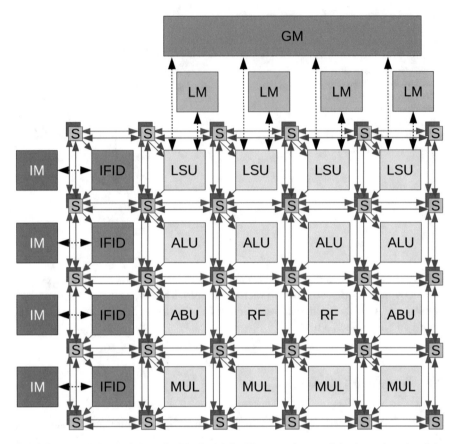

Fig. 3.8 Example instantiation of a Blocks fabric. The control network is shown in red and the data network is shown in blue

requests to the global memory within a clock cycle. To manage the requests an arbiter is placed between the global memory and the LSUs. When a load or store request arrives, the arbiter will check if these requests can be grouped (coalesced). For example, multiple reads to the same memory word can be coalesced into a single memory access. If multiple accesses are required, the arbiter will stall the compute fabric until all data has been collected. Although stalling the computation degrades performance, it is required to keep the memory accesses in sync with the operations in the compute fabric. Alternatively, a GPU-like approach can be used where fast thread switching is used to hide memory latency. However, Blocks allows the construction of deep compute pipelines that have many intermediate results stored in FU output registers. A context switch would require swapping all these values with intermediate states of another thread. This causes significant area and power overhead. In many cases, especially for DSP algorithms, stalls can be avoided by taking care of the memory access patterns. The arbiter in Blocks operates on a

Fig. 3.9 The Blocks memory
hierarchy consists of two
levels: the lower level
consists of local scratch-pad
memories that are private to
each LSU, and the second
level is a global memory that
can be accessed over a shared
bus. Access to this bus is
orchestrated by an arbiter.
Register files present in the
Blocks fabric can be
incorporated as an extra layer
of the memory hierarchy
close to the function units

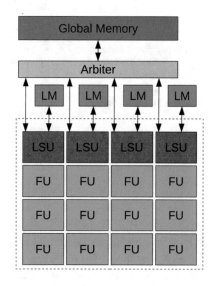

round-robin basis, ensuring that each request is handled within a bounded number
of cycles. Figure 3.9 shows an example of a Blocks architecture instance with four
LSUs connected to an arbiter.

In many applications data, or intermediate results, can be reused for multiple
computation cycles. Take for example a convolution; in such an operation all
elements in a window are multiplied with a signal. Both the signal and the
convolution window can be multi-dimensional, but for simplicity a one-dimensional
window and signal are assumed for the example shown in Fig. 3.10. In the first
iteration all data elements that are to be multiplied with the window coefficients
have to be loaded from memory, for example the external memory (indicated by the
blue squares). The multiplication is then performed; all products are added and the
result is stored in memory (indicated by the red square). In the next iteration, the
naive approach would be to load all required data again from the external memory.
However, this is not required since most of the values used in the first iteration can be
reused in the second iteration. Therefore, only one access to the expensive external
memory is required instead of five accesses. This is known as data reuse.

To perform data reuse the data that already has been loaded from the external
memory has to be stored somewhere. In the example these are only four values but
in reality, especially for multi-dimensional convolutions, the amount of data that
can be reused is often much larger. To efficiently exploit this, a memory close to the
compute units is required. Scratch-pad memories or caches are usually applied to
exploit data locality. Blocks incorporates scratch-pad memories close to the load–
store units. These are preferred above caches because it gives the programmer
direct control over the data reuse, thereby enabling specific reuse patterns to be
efficiently exploited. Additionally, they are more energy efficient than caches.
The local memories are private to an LSU, meaning that only the LSU that it is
connected to can access it. Although this distributes results across the memories,

Fig. 3.10 Example of one-dimensional convolution. Blue represents data to be loaded, orange indicates data that can be reused from a previous iteration, and red indicates produced data

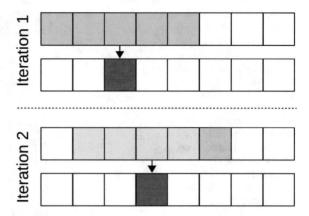

it has the advantage that no arbiter is required. Therefore, no stalls can be caused by accessing the global memories. Providing each LSU with its own local memory also has the advantage that when LSUs are operated in an SIMD manner, the total bandwidth increases linearly with the number of LSUs, thus not slowing down vector operations.

3.4   Configuration

Before Blocks can execute an application, it is required to configure the fabric with the design provided by the user. This design consists of two parts: a static configuration for the switch-boxes and the compute fabric (such as the function unit type configuration of instruction decoders), and a set of instructions for each instruction decoder that is used. Both of these parts are specified in a binary, described in more detail in Sect. 4.5, which is generated by the Blocks tool-flow.

There are several options to load such a binary into the CGRA fabric. For example, the configuration could be completely loaded under control of a host processor. The host processor then takes care of writing parts of the binary into specific memory regions that are made available to it. This has a few disadvantages: first, for each different type of host processor a new implementation has to be made for the configuration software; second, allowing direct access to the configuration can cause accidental hardware configurations that could cause damage to the hardware. For this reason, Blocks uses a hardware boot-loader that can be configured through a standardized interface. This allows the host processor to place the binary in any memory that Blocks has access to. This can be the same memory as where data is located, such as shown in Fig. 3.3. A pointer to the location where the configuration resides in this memory is then passed to the boot-loader that is sufficient to start the configuration process.

Configuring Blocks is performed in two steps. First, the interconnect and static function unit configuration are loaded. This configuration is part of the binary and

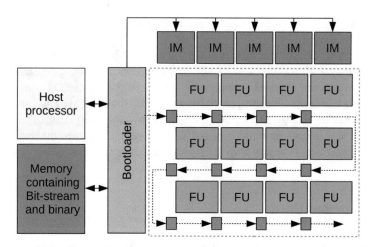

Fig. 3.11 The boot-loader within the Blocks platform is used to configure the networks and place the instructions in the corresponding instruction memories. These instruction memories are connected to special 'instruction decode' units within the reconfigurable fabric (not shown)

is stored as a bit-stream, a long sequence of bits that describes the configuration of various parts of the fabric. Each of the building blocks in the fabric that requires configuration has a register to hold the configuration values. These registers are connected in a long chain that functions as a single, long, shift register. Figure 3.11 shows an example of how this shift register is implemented in the fabric. On every configuration of clock cycle the boot-loader provides one bit of configuration data that is loaded into the shift register. This process is repeated until all configuration bits are loaded. It would be possible to replace the shift register with an addressable memory. The advantage of this would be that it allows partial reconfiguration of Blocks and allows faster configuration. However, the locations of the configurable bits of hardware are spread through the design. This may lead to longer paths from the configuration bits to this hardware. Investigating this trade-off is considered future work.

Once the bit-stream configuration is complete the second step in the boot-loader process has to be performed, which is loading the instruction memories. Every instruction decoder in the fabric has its own instruction memory attached. This is required to provide the required memory bandwidth to perform instructions for multiple instruction decoders in parallel. If all instructions would be loaded at run-time over a memory interface with a fixed width, for example 32 bit, stalls in computation would have to be inserted due to congestion on this bus when loading instructions. Additionally, instruction loading might interfere with loading and storing data. Separate instruction memories solve this problem but require the instructions in the binary to be copied to the instruction memories. The hardware boot-loader takes care of this operation. Once this process is completed, the binary located in the memory shared between the host processor and the Blocks is no longer required and may be overwritten, for example, with the next kernel that has to be executed by Blocks.

Execution of Blocks can be controlled via the boot-loader. It is possible to reset and reconfigure Blocks as well as obtain its execution status. For example, the status registers can be used to determine whether configuration is complete and successful, whether execution has finished, and if Blocks is stalling for memory accesses. Additionally, it is possible to obtain some hardware performance counter values. These indicate the number of stalls, memory accesses, and the total number of cycles executed. This may help in application debugging on a real device.

3.5   Function Units

Function units in CGRAs come in all shapes and sizes. They range from completely fixed function units that only perform a single (configured), such as those in systolic arrays, to miniature processors found in CGRAs close to many-core processors. Statically configured function units only require configuration bits once per application and, therefore, have minimal energy overhead. An example of such a function unit is shown in Fig. 3.12a, and this function unit is configured to perform an addition for the duration of the application.

Function units that represent small processors have a high energy overhead as they require more extensive, per cycle, reconfiguration via instructions. Figure 3.12b shows an example of such a function unit. This function unit contains an ALU, capable of performing arithmetic, logic, and shifting operations, as well as a multiplier unit and a register file for storing intermediate values. These operations are controlled by an instruction. Not all parts of such a function unit are always used in every cycle, but they still contribute to the total energy of the function unit. For example, when an addition is performed that can directly be forwarded to another function unit, both the register file and the multiplier are unused. However, this function unit is much more flexible. Due to its flexibility it can perform hardware

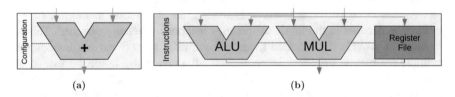

(a)   (b)

Fig. 3.12 Two types of function units found in CGRAs. (**a**) A statically configured function unit, the unit in this example is configured to perform additions for the duration of the application. (**b**) A flexible function unit. This function unit can perform cycle-based operations that involve ALU operations (arithmetic, logic and shifting) as well as multiplication and register operations. Operations are controlled by instructions

reuse in the temporal domain. Where the fixed function unit can only be enabled or disabled, the more complex unit can change the operation it is performing.

Multiplication and register file operations consume more energy than additions or logic operations. However, in the more complex function unit shown in Fig. 3.12b changes of the input values will ripple through both the ALU and the multiplier, as well as the multiplexing logic of the register files. This can be avoided by either inserting isolation logic in front of these units or placing input registers before the ALU, multiplier, and register file. However, both these methods have disadvantages. Operand isolation will increase the length and registering the inputs will cause additional cycle latency and in general higher energy. It makes more sense to divide the capabilities over specialized, heterogeneous function units. This prevents unwanted rippling effects through expensive, unused, parts of a larger function unit without isolation or input registering, while at the same time allowing these units to be used in parallel. An additional advantage is that specializing function units, and thereby reducing the number of different types of operations they perform, reduces the required instruction width. Specialized function units have another advantage: when care is taken in determining the instruction encoding, it is possible to reuse encoding patterns over function units. For example, an addition can use the same bit pattern as a multiplication since these operations are performed on different function units. For the instruction decoders this means that depending on their configuration (which type of function unit they control) the decoded instruction can be different although the same instruction is decoded. Of course, if possible, the decoded instructions should be as similar as possible to reduce the complexity of the instruction decoder. The reduced instruction width leads to smaller instruction memories, saving energy. Table 3.1 shows an overview of the function units currently available in Blocks. Although specialized function units

Table 3.1 Summary of function unit properties

FU type	Description	Supported operations
ALU	Provides typical RISC operations. One of the ALU outputs can be used unbuffered, allowing operation chaining	Arithmetic, logic, comparison, shifting
ABU	Can be configured in two modes, based on the bit-stream. Accumulate mode uses the ABU as a 16-register accumulator. Branch mode generates program counter values	(Conditional) branching, accumulation
LSU	Performs memory operations to global data memory and local memory, including configurable address generation	Global memory load and store, local memory load and store
MUL	Performs multiply operation that can be combined with arithmetic shift-right operations for efficient computation of fixed point results	Multiplication, shifting
RF	Provides 16-entry register file support to Blocks. RF operations in Blocks are explicit	Reading/writing registers
IU	Generate constant values onto the data network	Immediate generation

have a lower internal complexity, they require more connectivity to the network. This does increase the network overhead, but the expectation is that the increased parallelism leads to a performance increase that will cancel out the energy overhead of the network.

For this reason, Blocks uses moderately specialized function units. There are different function units that provide arithmetic and logic operations, multiplier and shifting, accumulation and branching, memory access, and local variable storage. Apart from the register file itself, the function units only contain registers on their outputs. This allows results to be directly communicated between function units, over the data network. Such direct communication reduces memory accesses significantly. As memory accesses are much more expensive in energy than computation, reduction thereof will lead to large energy reductions.

Most function units have multiple input ports and output registers connected to the data network, which are selected by the instruction; see Fig. 3.13. These selectable input ports serve as a small part of the interconnect that can be reconfigured (selected) on a cycle basis. The instruction is provided by the instruction decoder connected to the function unit over the control network.

The number of inputs and outputs is a design parameter that can be specified per function unit and thus can be changed by the designer depending on what the set of applications running on Blocks requires. By default, the function unit template instantiates four inputs and two outputs. The use of four inputs provides sufficient flexibility for most applications as it allows selection of two operands from other function units, one from a self-loop (loop back from the output to one of the inputs) and another input for connection to an immediate unit for constants. The use of two output buffers allows two values to be processed without requiring intermediate storage (for example in a register file) of one of these values. Such a situation occurs often in ALUs that are used to compute loop boundaries. For example, the current loop counter is stored in one output buffer for each loop bound computation and returned via a self-loop to be used as an input. This value can then be compared with a constant (e.g. provided by an immediate unit), and the result of this comparison can be provided on the second output to continue or break the loop. As writing to

Fig. 3.13 Instantiation of the function unit template, configured with four data inputs and two data outputs. The control network connection is marked with 'instr'. Some units, like the ALU, can be configured to bypass output registers

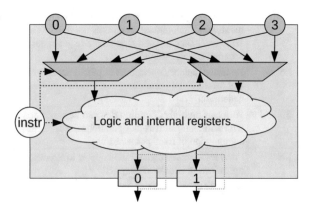

the second output will not affect the first output, the loop counter can be reused. A similar construction can be used for situations where a constant coefficient has to be multiplied with data values, common in various filters. The coefficient can be stored in one of the output registers and reused over computations.

Values are buffered at the FU outputs and not at the FU inputs, since there are generally fewer outputs than inputs, thus reducing the overall number of register bits required. The advantage of providing registers at the inputs instead of the outputs is that when input values toggle (due to upstream computations in the configured pipeline) these for not cause a ripple through the logic in a next function unit. However, since input registering requires more bits per function unit, and these ripples can be controlled by carefully designing the instruction schedule, output registering was chosen. A detailed analysis of this choice could be performed as future work.

Appendix A describes each type of function units in more detail. The Blocks framework allows easy extension with new function unit types. To do so, the new function unit itself has to be described in a hardware description language. This description is added to the function unit templates together with a description of the instruction encoding that is to be performed by the instruction decoders. The instruction encoding and template system is described in Chap. 4.

3.6 Construction of Architecture Instances

With its flexibility Blocks allows various architecture types to be specified. These structures can be very similar to existing processors, such as VLIWs and SIMDs. However, they can also be completely adapted to the application and provide spatial layout for (part of) the application. The construction of VLIW processors is described in Sect. 3.6.1. The construction of SIMDs is described in Sect. 3.6.2. Section 3.6.3 concludes with a description of how application specific data-paths can be constructed.

3.6.1 Constructing VLIW Processors

VLIW processors execute instructions that specify multiple operations. These operations control multiple issue slots at the same clock cycle. Since not all operations are required to be the same, it is possible to use VLIWs for ILP. Instead of using a single instruction fetch and decode unit that loads a very wide instruction, Blocks uses multiple instruction fetch and decode units that are connected to the same program counter. To do so, each IFID has a connection to the data network to connect to the output of an ABU, which produces a program counter value. Figure 3.14 shows an example of such a structure for two IFIDs. By doing so it comes under control of the sequencing provided by the ABU. Since the program

Fig. 3.14 Construction of a
2-issue VLIW processor
using the Blocks fabric. The
program counter (PC) is
distributed to multiple
instruction decoders to make
these operate in lock-step and
form a VLIW processor

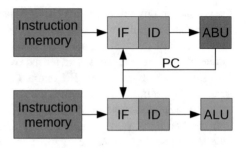

counter is available on the network, every IFID can connect to the program counter.
As the program counter is the same for every connected instruction decoder all
decoders will run in lock-step, meaning they follow the same sequence. Each
instruction decoder however is connected to its own instruction memory. These
memories, prepared by the boot-loader, contain the instructions that need to be
executed. From an architectural point of view this allows the instructions at the
same addresses to be 'concatenated'. In essence, the individual narrower instruction
memories operate as a single wide memory producing a wide instruction. This is
also what a VLIW processor does, and it loads very wide instructions that control
multiple issue slots at once.

The advantage of the Blocks method is that the width of the VLIW processor
(i.e. the number of issue slots) can be easily matched to the requirements of the
applications. When a wider processor is needed, more instruction decoders can be
connected to the program counter generated by the ABU, as shown in Fig. 3.14.
If fewer issue slots are required, some instruction decoders can be left unconnected, reducing power. The disadvantage of the flexible processor construction
that Blocks uses is that some VLIW optimizations cannot easily be applied. For
example, VLIW processors can reduce some of their power by applying instruction
compression. When issue slots are unused, their operations are not specified and
the instruction width is reduced. This saves some unnecessary decoding that saves
power. In contrast, Blocks stores instructions for each instruction decoder in separate
instruction memories. Furthermore, which instruction decoders are going to be used
in a specific configuration is unknown at design time. Therefore, it is not possible
to compress over multiple instruction decoders; this limits the use of instruction
compression significantly. One possible solution for this is to cluster instruction
decoders within the Blocks framework. For example: four instruction decoders can
be merged into a single wider instruction memory where instruction compression
can be applied. This has some consequences for mapping which instruction decoders
will be allocated for a specific function and reduces some of the flexibility of Blocks.

Since the flexibility of Blocks allows the architecture to adapt to the application,
it is often possible to apply software pipelining. The idea behind software pipelining
is to overlap different iterations of loop bodies in time; i.e., the next iteration
starts, while the current iteration still has to finish. This shortens the effective
iteration time. In the extreme case, the iteration interval becomes a single cycle.

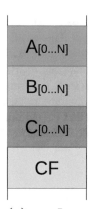
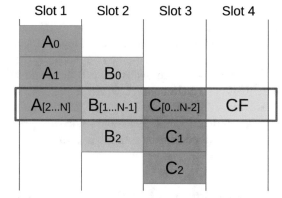

(a) Loop execution on a traditional processor.

(b) Loop execution using software pipelining in Blocks. An inital prologue and eventual epilogue of the loop are used to start and end loop iterations, during loop execution however the instruction is static. This also includes the control flow.

Fig. 3.15 Loop execution schemes for a traditional single issue-slot microprocessor (**a**) and a software pipelined Blocks instance (**b**)

Figure 3.15b shows an example explaining software pipelining. With software pipelining instruction-level parallelism is exploited by spatially mapping the operations, often of a loop body, over multiple function units. The number of function units required, and the connections between these units, depends on the content of the loop body that is mapped onto the hardware. Blocks can, as long as there are sufficient resources, efficiently realize this by instantiating a VLIW processor with the right number of issue slot and bypass connections. Computation of the loop bounds can also be part of the software pipelining , allowing loop bodies and their overhead to be reduced to a single, static, instruction. This instruction can then be constant for the duration of the loop-nest, significantly reducing instruction fetching and decoding power. Figure 3.15a shows loop execution for a traditional microprocessor. In such a processor all operations are performed sequentially; this leads to a data-path that is reconfigured every cycle. To do so, a new instruction needs to be fetched and decoded on a cycle basis. In contrast, Blocks can implement a VLIW processor that matches the required issue-slot width and perform software pipelining. This allows the loop execution to consist of a single, static, operation. An example of such a schedule is shown in Fig. 3.15b.

The global memory is managed by an arbiter that decides which LSUs get access in a specific clock cycle. It may occur that not all requests from all LSUs can be completed at the same time. In this case the virtual processor realized on the Blocks fabric stalls to wait for the memory operations to complete. It is important to note that when multiple processors are instantiated on the Blocks fabric that only those affected by the memory operations will stall. Stalls are initiated by the LSUs; as long as a request is outstanding a 'stall flag' is raised. The stall flags of all LSUs that are part of the same processor are or-ed together and forwarded to the ABU and IDs. The ABU then stalls execution by not updating the program counter. Multiple of these or-ed networks are available, and the static configuration determines to which network an LSU connects. When a stall occurs, the instruction decoders disable writing of all output buffers on all function units such that the current state is maintained during the stall. Synchronization between multiple instantiated virtual processors is described in more detail in Sect. 3.7.

3.6.2 Constructing SIMD Processors

Typical SIMD processors use one instruction decoder to control multiple vector lanes, as shown in Fig. 3.16a. Each vector lane has a register file, ALU, multiplier, and an LSU. In most cases an instruction controls the register file and one of the three other function units. For example, to perform an addition the register file and ALU are controlled simultaneously. Therefore, only one of the function units (other than the RF) is active at one time. All units are controlled by one, somewhat wider, instruction. This means that if an application has to be executed that performs a memory load, an addition, a multiplication, and a memory store, then this will take (at least) four cycles. Each cycle another instruction is issued from the instruction decoder.

The SIMD processors constructed on the Blocks fabric differ from typical SIMD processors. An architecture instance for an example application is given in Fig. 3.16b. Since the function units are specialized and controlled by their own instruction decoder, the Blocks approach results in a VLIW-SIMD processor. This means that the vector lanes for each function unit type effectively form an issue slot of a VLIW processor. Although the total number of instruction bits is wider than that in the typical SIMD processor (in this example, 48-bit for Blocks versus

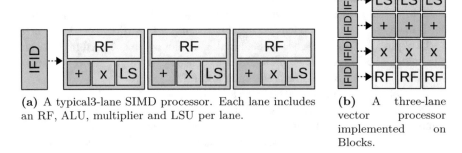

(a) A typical 3-lane SIMD processor. Each lane includes an RF, ALU, multiplier and LSU per lane.

(b) A three-lane vector processor implemented on Blocks.

Fig. 3.16 SIMD processors, the typical approach (**a**) and the Blocks approach (**b**)

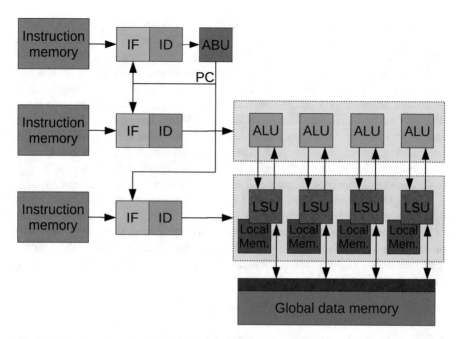

Fig. 3.17 Construction of an SIMD processor using the Blocks fabric. The ALU and LSU function units are controlled as a vector lane of four elements wide

typically 32-bit in the SIMD) the instruction for the example application can be software pipelined with an iteration interval of a single cycle. This reduces power in the instruction decoders and their attached instruction memories due to reduced toggling. Additionally, the Blocks approach will provide a higher throughput since the operations can be pipelined.

A more detailed example of supporting SIMD lanes by Blocks is shown in Fig. 3.17. SIMD significantly reduces the instruction overhead compared to the local or global decoding used by traditional CGRAs. Construction of vector lanes is very similar to constructing a VLIW issue slot as described in Sect. 3.6.1. An ABU is used to control the sequencing of an instruction decoder. This instruction decoder then loads and decodes the corresponding instructions. The decoded instructions are made available on the control network and routed to multiple function units. Since these units all receive the same decoded instruction, they will perform the same operation in the same cycle. The inputs and outputs of the function units of these function units are connected to different sources and sinks, such that they operate on different data. This data can, for example, be provided by LSUs, which also operate as a vector lane, or by other function units.

Since the inputs and outputs of the function units that operate under control of the same instruction decoder can be connected to any other function unit on the Blocks fabric, very interesting constructions can be made. For example, an application that performs both horizontal and vertical projection on an image sums all rows and columns in an image into two 1-dimensional vectors. Since both operations involve summing values, at the same cycles, both the ALUs for vertical and horizontal summation can operate under the control of a single instruction decoder. The different data sources determine what is actually computed. Of course, this is just an example but it shows that vector operations can be used in multiple ways.

Blocks allows direct communication between FUs, without using a register file. This allows intra-SIMD lane, but also inter-SIMD lane, communication. This enables shuffling or shifting operations between vector lanes in fixed patterns, which are useful in applications like the Fast Fourier transform (FFT) and various filters.

To supply data to the vector lanes, LSUs can be operated like a vector lane as well. Each LSU can provide read and write access to a private local memory and a shared global memory. Currently, the Blocks applications need to explicitly load data from global memory and store it in the local memories. For future versions of Blocks, changes to the memory hierarchy are envisioned where either the host processor or a Direct memory access (DMA) controller can directly write into the local memories. This will allow for parallel data loading and execution, improving the function unit utilization.

3.6.3 Application Specific Data-Paths

Besides the typical processor structures, it is possible to create very application specific designs. Doing so can significantly improve the performance or energy efficiency of the application as this allows Blocks to create (near) spatial layout of applications. In some cases, part, or whole, of these special structures can be controlled with a single instruction decoder. For example, when multiple values have to be added together, an adder tree can be constructed. An example thereof is shown in Fig. 3.18. The whole tree can be controlled by a single instruction decoder when the application is in streaming mode, acting as a two-dimensional SIMD instruction. A similar structure can be used for other reduction trees, such as maximum, or minimum, searching. In fixed processors these structures are sometimes added as an optimization or accelerator. However, they are only used when the application can be properly mapped to these accelerators. In Blocks, they can be instantiated when, and with the properties, required.

Blocks allows some function units to be chained, i.e. to be interconnected with a bypass for the output register. The result of this is that the output is unbuffered, allowing the construction of multi-unit single-cycle operations. This should, of course, only be done when it causes no increase to the critical timing path. For example, when two ALUs are 'chained' in a single cycle in parallel with a multiplier operation (single cycle), no increase in the critical path delay is expected.

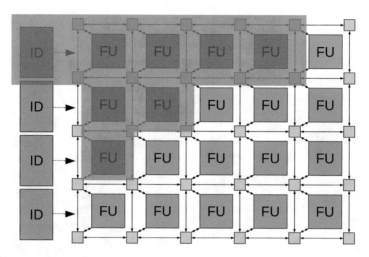

Fig. 3.18 Construction of application specific structures. This figure shows an example of a reduction tree that can be used for summation, maximum finding, etc.

3.7   Multi-Processor, Communication, and Synchronization

There is no restriction on the mix of function units that can be placed onto the Blocks fabric at design time. Therefore, multiple ABUs can be present, where each can generate their own program counter. Having multiple program counters allows for construction of multiple, independent, processors onto the same Blocks fabric. These processors will of course run from the same clock, but their instruction sequencing is independently controlled by their own ABU. This can be achieved by routing the respective program counters to different sets of instruction decoders. The individual sets of instruction decoders then operate in lock-step and form their own VLIW-like processor; Fig. 3.19 shows an example of two VLIW processors configured on the Blocks fabric.

Section 3.7.1 describes how synchronization in Blocks can be achieved based on global memory locking. Section 3.7.2 introduces local memory sharing to enable communication between instantiated processors. Section 3.7.3 builds on this concept with inter-LSU communication. Section 3.7.4 describes how a DMA can be used to offload global memory to local memory data transfers to a DMA unit. A lightweight, First-in first-out (FIFO) based, method to enable value passing and synchronization between processors is elaborated in Sect. 3.7.5. Finally, Sect. 3.7.6 describes a method to hide global memory latency with a form of multi-threading [5].

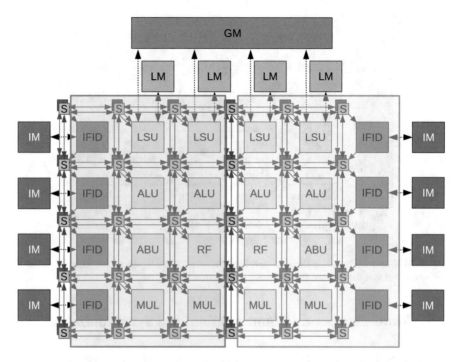

Fig. 3.19 Two independent processors on the Blocks fabric (shown in green and yellow). By using two ABUs, two independent program counters can be generated. IDs are controlled by one of these program counters, depending on the network configuration

3.7.1   Locking on Global Memory

Multi-processor configurations lead to an interesting problem with respect to storing and retrieving data from global memory. As mentioned, it is possible that the arbiter requires more than one cycle to complete the global memory operations. If this is the case, the processor needs to be stalled until all data are available. However, this cannot be a Blocks-wide (global) stall as there is no need to stall a processor that did not perform a memory operation. This processor can continue, independently of the other processors realized on Blocks. To overcome this, Blocks has multiple 'stall groups'. Each instruction decoder, LSU, and ABU can be connected to one of these stall groups; this is specified in the bit-stream and, therefore, performed at the start of a kernel or application. The number of stall groups is a design parameter and determines the maximum number of individual processors at run-time.

In many cases the individual processors run independent tasks that only synchronize occasionally, they are required to operate as a multi-processor platform. This involves communication and, even more importantly, synchronization between the processor cores. Several methods for inter-processor communications exist; examples thereof are shared memories, semaphores, and sockets. Blocks supports communication via the shared global memory. In [5] this concept is explained in more detail.

In this case semaphores are used that synchronize the processors and maintain consistency. Although this method does not require any additional hardware it requires software implementation of the synchronization methods. This may lead to inconsistencies or even blocked processors when not done correctly. Furthermore, multiple read and write operations are required to implement the semaphore. This results in higher memory bus traffic and may introduce stalls on the bus. All these add up to a relatively high cycle count per synchronization event.

3.7.2 Local Memory Sharing

There are various alternatives for synchronization and inter-processor communication. For example, LSUs could share access to their private local memory with one or two neighbouring LSUs. The disadvantage of this method is that it then becomes necessary to either have multiple ports on these memories or to perform arbitration between the LSUs using it. This is not efficient in both area and energy. Figure 3.20 shows how such a structure could be implemented. The blue arrows show the necessary extension to the Blocks fabric.

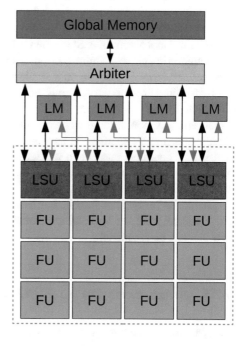

Fig. 3.20 Local memory sharing between two LSUs. Since this requires extra ports on the memory, this method is not efficient, neither in area nor in power and performance

3.7.3   Direct Inter-LSU Communication

Another option would be to allow LSUs to directly communicate between each
other. To implement this the arbiter could consider other LSUs as a data source and
allow access to their local memory through the LSU. By performing data movement
through the LSU it is not required to use multi-port or arbited local memories for this
purpose. This can be implemented as an arbited bus, as a crossbar network or even as
a network-on-chip (NoC). A bus-based implementation is smaller and consumes less
power but has the disadvantage that only one link between two LSUs can be active
at the same time. This may be sufficient for communication and synchronization
between two processors. However, it is unsuitable for vector loads unless these can
come from one source LSU and requests can be coalesced into a single access. A
crossbar implementation and a NoC will provide multiple simultaneous connections
and, therefore, vector based communication, but at the expense of area and power.
A NoC may provide variable latency depending on the path that has to be taken.

3.7.4   DMA Controller

A DMA controller, capable of moving data between local memories, would be
another effective solution. The DMA can operate independently from the processor
fabric and overlap memory operations with computation. An additional advantage
of such a method is that data can be placed in exactly those local memories where
it is required at a given time. Since the DMA requires access to the local memories,
these memories either require two ports for fully independent DMA and CGRA
accesses or use cycle stealing on a single port. When the CGRA compute fabric is
not performing any memory operations, the unused memory cycles can be used by
the DMA for data movement. The disadvantage of the latter method is of course
a reduced, and more unpredictable, bandwidth. However, a two-ported memory is
significantly more expensive in area and power than a single ported memory. The
DMA itself could operate in two ways: either it can have its own off-line scheduled
operations that can be determined by the programmer or compiler or LSUs can
access configuration registers in the DMA controller to set up transfer of data blocks.
The latter option is more flexible but leads to a more complex DMA and requires
active control from the application running on the Blocks fabric. Figure 3.21 shows
how this can be organized.

3.7.5   Synchronization via FIFO

For synchronization purposes a FIFO can be added in the memory subsystem. This
FIFO can be memory mapped on the global (shared) memory bus, allowing all LSUs
to access this FIFO. An example thereof is shown in Fig. 3.22. Just like accesses to

Fig. 3.21 Local memory data management using a DMA. The DMA has access to the local memories (indicated with the blue arrows), either via a second port or by cycle stealing on the port that connects to the LSU. The DMA can be controlled using configuration registers that are accessible as a memory mapped device (indicated with the black arrow from arbiter to DMA)

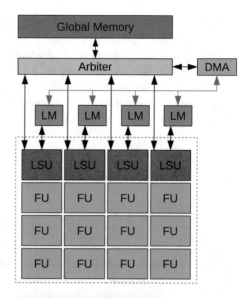

Fig. 3.22 A FIFO added as a memory mapped device, the FIFO can be used to pass values from one processor to another or be used to synchronize two processes. Multiple FIFOs could be added if desirable

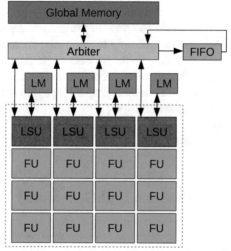

the global memory, accesses to the FIFO will be arbited and can even be coalesced. When an LSU writes to the FIFO, the data will be placed in the first available data element and be available for reading by another LSU. When the FIFO is full, subsequent write attempts to the FIFO will be stalled until a read is performed on the FIFO. Reading from the FIFO frees up a data element; when all data elements are read, any subsequent reads will be stalled until data becomes available again. The stalling behaviour on the read and write side of the FIFO can be used to synchronize processors without using explicit software synchronization. The data elements in the

FIFO can either be 'tokens' (with no actual meaning other than synchronization) or actual results that need to be communicated.

Overall, the best solution to allow multi-processor communication and synchronization within Blocks seems to be a combination of a memory mapped FIFO for synchronization and a request based DMA controller for data movement. This setup requires the smallest modifications to the existing architecture and the highest degree of flexibility.

3.7.6   Hiding Memory Latency

To hide global memory latency, Blocks can be configured to incorporate a form of context switching. To do so a special module called the 'state controller' is instantiated. The state controller has access to a scan-chain of shadow registers that are located besides all primary registers inside the data-path. At any point during execution the program flow can be paused and the contents of the registers can be swapped with the contents of the shadow registers. This effectively implements a context switch as the whole data-path (including program counters) now changes to another state. The state in the shadow registers can either stay there until a future context switch (if only two contexts are required) or be scanned out into one of the memories that Blocks have access to (if more than two contexts are required). Any context stored in this memory can be reloaded at any time. Operation of the context switch module is under explicit operation of the host processor and is controlled via configuration registers. This feature allows a simple form of multi-threading but is relatively expensive in area as it duplicates all registers; therefore, it is switched off by default.

3.8   Function Unit Granularity

Different applications have different requirements on the width of the data-path. For example, some applications only require 8-bit data, while others operate on a mixed data-width. Blocks allows design-time selection of the data-path width; the hardware description will then be generated for the specified data-width. The available data-width configurations that can currently be selected are 8-bit, 16-bit, 32-bit, and 64-bit. Fixed configurations, however, leave room for improvement as dynamically matching the data-width of the application to the architecture prevents unnecessary computation and toggling of signal lines. Therefore, supporting adaptive data-width has been investigated by exploring scalable function unit granularity [6].

The data-width of a function unit determines the granularity of the Blocks fabric. Therefore, function units that support multiple data-widths efficiently are multi-granular. Within the Blocks fabric two types of function units are of interest when considering support for multi-granular operations: the ALU and the multiplier units. Multi-granularity can be achieved in two ways: smaller function units can be combined into larger function units, or larger function units can internally be split into smaller function units. Both methods have their advantages and disadvantages. For example, combining function units requires more interconnect routing while splitting function units is less flexible. The best solution is application dependent.

For addition, the modifications that have to be made to function units to either support splitting or merging are very similar. For example, a 32-bit ripple adder can be split into four 8-bit adders by cutting the carry signal in the appropriate location. Since the carry now does not propagate any more between the 8-bit adder segments, these function as independent vector adders. Merging four 8-bit adders into a single 32-bit adders is very similar; instead of cutting the carry within a single unit we allow the carry signals to be propagated (e.g. over a dedicated network) between function units. This effectively combines the multiple units into a single, wide, adder. Figure 3.23 shows how multiple ALUs can be merged into a wider ALU, as well as how a wide ALU can be split into multiple narrower vector ALUs. The multiplexers that determine ALU granularity are controlled by special instructions. The network connection that determines how the carry-chains are built is controlled by the Blocks bit-stream. The current support for multi-granular units requires a carry-in source to be specified in the architecture description. This connection will be instantiated as a fixed wire between these units to reduce overhead. An internal multiplexer in an ALU determines whether the connection is used or not.

When it is possible to perform multi-granular, addition multi-granular multi-plication becomes also available. Multiplications can be performed by performing parts of the multiplication (partial products) on smaller function units. These partial products are then summed to produce the final result, which can be performed using multi-granular adders. The Blocks fabric allows this kind of multi-granular computations by configuring the network around the function units to form the right

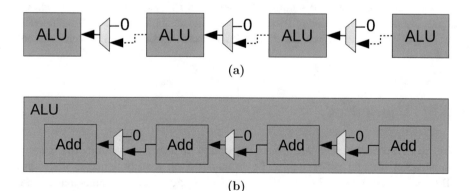

(a)

(b)

Fig. 3.23 Splitting and merging adders into narrower or wider function units to adapt to the required data granularity. (**a**) Merging multiple ALU function units into a wider ALU. The dotted line shows the carry result transmitted over a special carry network between function units. (**b**) Splitting adders inside one ALU into multiple narrower vector adders

structure to perform these merged operations. An additional advantage of such a
method is that multiplications with operands of different widths can be supported.
For example, it is possible to form an 8 by 32-bit multiplier out of 8-bit multiplier
blocks. In a fixed architecture such a multiplication would be performed on a 32 by
32-bit multiplier.

3.9 Approximate Computing

Approximate computing is a topic on which a lot of research is performed in recent
years. It allows inexact computations to be performed. Such computations can have
a limited error in the computed result. Approximate computing aims at a lower
energy per computation, smaller hardware, or higher performance due to a reduced
critical path length. Approximate computing is especially popular in the field of
machine learning. Neural networks often do not require calculations to be fully exact
in order to provide results that still yield acceptable precision. Another situation in
which approximate computations can be used is where the results produce visual
content, such as video. In many cases the human eye will not discern any inaccuracy
in the image produced. Of course, the program flow still needs to be computed on
fully accurate function units.

The level of approximation that can be allowed is highly application dependent.
It is, therefore, interesting for function units to provide variable levels of approximation. Also in this case, the flexibility of Blocks allows function units to do so.
Existing solutions usually have function units that provide a configurable level of
approximation. Blocks only requires the function units to be configurable between
either accurate or approximate to some degree. Essentially, approximation can only
be switched on or off. By applying this to narrower function units (e.g. 8-bit) it
is possible to use the multi-granularity support in Blocks to form wider function
units. By configuring some of these function units as approximate, a suitable level
of approximation can be reached. Figure 3.24 shows an example on how this can be
achieved.

3.10 Conclusions

The Blocks concept revolves around separation of data and control. This allows
application specific processors to be instantiated on a reconfigurable platform.
Blocks has, like most CGRAs, a two-dimensional grid of function units that can be
connected over a network. In addition to this, Blocks has separate instruction fetch
and instruction decoder units that connect over a separate network to one or more
function units. These IFIDs provide the control for the connected units. If multiple
function units are connected to the same IFID, these units operate in parallel as an
SIMD. Program flow is controlled by a special unit, the ABU. This unit generates a

Fig. 3.24 Using
multi-granularity within the
Blocks framework to provide
variable levels of
approximation. In this
example, three accurate
ALUs (blue) are combined
with one approximate ALU
(green). The approximate
ALU is used for the least
significant bits. The dotted
lines illustrate the connected
carry-chains between ALUs

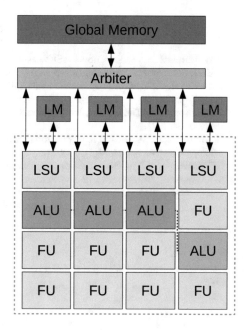

program counter that can be distributed, over the network, to one or more IFIDs to
allow construction of VLIW processor structures.

Blocks uses FPGA-like switch-box based networks. These networks are configured once per application or kernel and are static for the remainder thereof. In
contrast to the networks used in FPGAs, the networks in Blocks are bus-based
instead of providing only a single wire connection. This reduces the energy overhead
of the network.

The memory hierarchy of Blocks consists of a large global memory that can be
accessed by LSUs over an arbited interface. In addition to this, each LSU has a local
memory that is private to that LSU. This memory is generally used for data reuse.
The data loaded by the LSUs is provided onto the data network and can be used by
any of the other types of function units.

The flexible nature of Blocks allows construction of SIMD and VLIW processors
and a mix between these. It also allows construction of application specific data-
paths and multi-processor systems on the same fabric. A multi-processor system is
achieved by using multiple ABUs to generate a unique program counter for each
processor. These processors may occasionally have to synchronize or exchange
results. Blocks allows several methods to do so, ranging from communicating via
the shared global memory to special FIFO interfaces.

References

1. R.W. Hartenstein, et al., A novel ASIC design approach based on a new machine paradigm. IEEE J. Solid State Circuits **26**(7), 975–989 (1991)
2. B. Mei, et al., ADRES: An architecture with tightly coupled VLIW processor and coarse-grained reconfigurable matrix. *Field Programmable Logic and Application* (Springer, 2003)
3. S. Swanson, et al., The WaveScalar architecture. ACM Trans. Comput. Syst. **25**(2), 1–54 (2007). ISSN: 0734-2071
4. H. Dutta, et al., A holistic approach for tightly coupled reconfigurable parallel processors. Microprocess. Microsyst. **33**(1), 53–62 (2009). ISSN: 0141-9331
5. S.F.M. Walstock, *Pipelining streaming applications on a multi-core CGRA* (2018)
6. S.T. Louwers, *Energy efficient multi-granular arithmetic in a coarse-grain reconfigurable architecture* (2016)

Chapter 4
The Blocks Framework

The Blocks framework is based on a set of design parameters and configuration files. The hardware for Block is generated based on a set of templates and generation tools. Furthermore, the Blocks tool-flow is able to generate configuration bit-files for the generated hardware as well as binaries that are required for program execution on Blocks. This chapter introduces the templates and tools to use Blocks. Section 4.1 introduces the tool-flow and the steps required to generate a Blocks architecture instance. Section 4.2 describes all design parameters that can be adjusted in the tool-flow. Section 4.3 continues with detailing how hardware generation is performed. Section 4.4 explains how Blocks can be programmed. Section 4.5 describes the steps the tool-flow performs to generate executable files for Blocks. Section 4.6 concludes this chapter.

4.1 Overview of the Blocks Tool-Flow

Figure 4.2 presents a high-level overview of the Blocks tool-flow. The tool-flow uses a two-step approach to generate a reconfigurable architecture and its configuration. The first step is to specify a physical architecture. This architecture describes which function units are physically present in the design as well as the properties of the data and control interconnect networks. In the second step, the physical hardware design is configured for a virtual architecture description. It specifies which function units are used and to which physical function units they are mapped. Furthermore, the virtual architecture describes the connections between function units, both on the data network and the control network. This information is used by a routing tool to generate network configurations. The advantage of this two-step approach is that besides mapping a virtual architecture to a physical architecture, it is also possible to map the virtual architecture directly on e.g. an FPGA. Doing so results in application specific processors; essentially these are hard-wired versions of Blocks

© The Author(s), under exclusive license to Springer Nature Switzerland AG 2022
M. Wijtvliet et al., *Blocks, Towards Energy-efficient, Coarse-grained Reconfigurable Architectures*, https://doi.org/10.1007/978-3-030-79774-4_4

Fig. 4.1 Applications for
Blocks are mapped onto a
virtual architecture. The
virtual architecture can be
mapped to a physical
architecture with a
reconfigurable interconnect.
It can also be directly mapped
to, for example, an FPGA

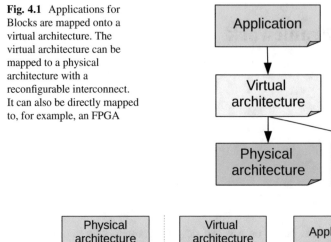

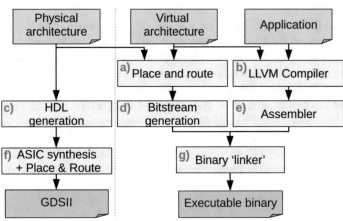

Fig. 4.2 The Blocks tool-flow. This tool-flow generates both the hardware and the executable
binaries for Blocks. The left part of the vertical dotted line typically has to be re-run only when
an architectural change has been made. The right part of the tool-flow is aimed at generating the
configuration and executable instruction files

where the interconnect is replaced by direct connections. This allows Blocks to be
mapped onto FPGAs for evaluation and debugging, as shown in Fig. 4.1.

The physical hardware description is used to generate Verilog that will be
synthesized (c), placed-and-routed (f), and eventually exported as GDSII. This tool-
flow path has to be executed only once for a specific Blocks instance. Only when
the design resources change, this path has to be re-run (Fig. 4.2).

The virtual architecture is used to perform automated placement and routing
on the Blocks fabric (a). Placement is performed using simulated annealing and
routing is performed using the PathFinder algorithm. The placed-and-routed design
is translated into a bit-file that is used to configure the Blocks fabric (d). An LLVM
compiler back-end (b) generates functional code, but for the benchmarks in this
chapter the assembly is hand optimized. Finally, the assembled program code and

bit-stream are merged into a binary format (g) supported by the Blocks hardware boot-loader.

4.2   Design Parameters

The Blocks framework has several design parameters that can be modified to make the generated architecture fit to the target application domain. Most of the architectural properties are described in configuration files. These configuration files are in the Extensible mark-up language (XML) format and are easy to modify. The structure of the configuration files is hierarchical, and there is a 'base' configuration file describing the main architectural properties and supported instructions. One level lower is another configuration file describing aspects such as the available function unit types. These files are for specific data-widths supported by Blocks. For example, there is a configuration file for the 32-bit variety and the 8-bit variety. The reason this is not simply a configurable parameter is that the number of input and output ports may vary for different widths. For example, the 8-bit version requires two 8-bit ports to connect the 16-bit program counter. Another level lower is the architecture file that describes the structure the user wants to implement. For example, the user can instantiate function units and connect these together and configure the switch-box networks. Each lower level file can override the properties of a higher level file. This allows the base architecture to specify, for example, a default memory size that can be overridden by the user if desired. Figure 4.3 shows the hierarchy of the described configuration files.

As an example, Listing 4.1 shows the instantiation of two function units for a Blocks architecture without switch-boxes (direct connections). Such a description is later used by an automatic place-and-route tool that implements the same function unit selection and their connections on a reconfigurable Blocks instance. The example listing shows two function unit instantiations, an LSU and an ALU. It can also be observed that these function units use each others' output ports. The ALU uses LSU output port 0 as the source for input 1, while the LSU uses ALU output port 1 as a source for input port 1.

```
base.xml  ........................describes common architecture properties
  8b.xml  ......................describes function unit properties for 8-bit
  16b.xml  ..................describes function unit properties for 16-bit
  32b.xml  ..................describes function unit properties for 32-bit
    architecture.xml  ...............user defined architecture instance
```

Fig. 4.3 File hierarchy of the Blocks configuration files

```
1  <fu type="LSU" name='lsu' ID="id_lsu">
2          <input index="0" source="imm.0"/>
3          <input index="1" source="alu.1"/>
4  </fu>
5
6  <fu type="ALU" name='alu' ID="id_alu" config="1">
7          <input index="0" source="imm.0"/>
8          <input index="1" source="lsu.0"/>
9  </fu>
```

Listing 4.1 Example of two function unit instantiations and connections between them

```
1  <functionalunittypes>
2          <ALU>
3                  <control idtype='2'/>
4                  <reconfiguration bits="1"/>
5                  <connections inputs="4" outputs="2"/>
6          </ALU>
7  </functionalunittypes>
```

Listing 4.2 Function unit type description example for an ALU

4.2.1 Function Unit Types

The number of input ports and output ports can vary per function unit type. In order to allow users to add their own function units to the Blocks framework the function unit types are described in an XML file. Since FU properties may depend on the data-width, separate XML specifications are needed for each required data-width. Listing 4.2 shows how the ALU is specified as a function unit type. As mentioned, the number of input and output connections is specified. In this case the function unit has four inputs and two outputs. For the ALU, one of these outputs can be configured to be either buffered or unbuffered. This is done using a bit in the bit-stream generated for the Blocks instance. Therefore, the function unit type is specified to have one reconfiguration bit. Each function unit type can be controlled by the same instruction decoder. However, these instruction decoders need to be configured to control a certain type of function unit. This configuration is performed at the start of the application. The 'idtype' field in the listing specifies which of the decoder configurations is required to correctly control the ALU; this is part of the static configuration. The configuration number effectively defines the highest (static per application) three bits used by the instruction decoders.

```
1  < instructiontypes >
2          < T1 >
3                  < mnemonic  type = " mnemonic " / >
4                  < output  type = " output " / >
5                  < inputB  type = " input " / >
6                  < inputA  type = " input " / >
7          </ T1 >
8  </ instructiontypes >
```

Listing 4.3 Example showing instruction type description

4.2.2 Instruction Specification

The base configuration file describes all instructions available for the function units. Instructions in Blocks are different from most architectures as the source and destination operands do not specify register locations. A Blocks instruction specifies which of the connections to the network (typically four) is used for each of the two operands. For example, the instruction specifies that the source for operand A is input port 1 of the function unit. The actual source of the operand value is thus determined by the network configuration, to which other function unit output it is connected. The destination of the result is specified as to which of the output buffers the result is written. These output buffers are available as sources on the network and can, therefore, be used by other function units as a source.

The instructions are categorized by type. An instruction type is a set of instructions that have the same parameters and width for these parameters. For example, an instruction with two source operands is a different type than an instruction without operands. Listing 4.3 shows the description for instruction type 'T1', and this instruction type has a mnemonic, a field specifying the destination output, and two fields to specify input connections for the operands A and B.

The field types are also described in the base configuration file. For example, the 'input' field is specified as a field that encodes to a value of two bits wide and that in assembly the argument format is starting with 'in' followed by a number. This allows the assembler to check that, for example, 'in1' is a valid argument for this field. The instructions can be specified by describing the mnemonic, the type, and the opcode. Listing 4.4 shows an example for some instructions of the ALU. It can be observed that the pass instruction, which only has a single input operand, is of another type than the remaining instructions that take two inputs for their operands. The numbers specified in the opcode fields are encoded into the instructions. This field specifies which operation has to be performed as a binary pattern.

```
1  <instructions>
2          <add mnemonic="add" type="T1" opcode="26"/>
3          <sub mnemonic="sub" type="T1" opcode="27"/>
4          <and mnemonic="and" type="T1" opcode="16"/>
5          <nand mnemonic="nand" type="T1" opcode="48"/>
6          <or mnemonic="or" type="T1" opcode="17"/>
7          <pass mnemonic="pass" type="T1_d" opcode="19"/>
8  </instructions>
```

Listing 4.4 Instruction specifications for the ALU

```
1  <Core>
2    <Memory type="TSMC">
3      <GM width="32" depth="4096" addresswidth="32" interface="DTL"/>
4      <LM width="32" depth="256" addresswidth="16"/>
5      <IM depth="256" addresswidth="16"/>
6    </Memory>
7  </Core>
```

Listing 4.5 Instruction specifications for the ALU

4.2.3 Memory Sizes and Interface

The memory hierarchy is an important part of the Blocks architecture. The required size and interface of the memories can be different for each instantiation target. For example, Blocks may need to be connected to a bus with shared memories instead of managing its own memories in a stand-alone fashion. Various aspects of the memory hierarchy of Blocks can be configured using the base configuration file. Of course, these properties can be overridden in the instantiation specific architecture descriptions, made by the user. An example of such a memory configuration is shown in Listing 4.5. In this example memory macros are used that are suitable for ASIC synthesis and place-and-route, specified using the 'TSMC' value. The interface is configured to use a device transaction level (DTL) bus to interface with the global memory and peripherals. Using the DTL bus allows Blocks to be connected to various other platforms such as the CompSoC platform [1].

The configuration files also allow specification of peripherals. Peripherals are typically modules that interface to the outside world, for example a serial interface or a display driver. These peripherals are memory mapped; i.e. they are placed on the same memory interface as the global memory. Therefore, LSUs can access peripherals in the same way as they would access the global memory.

4.2.4   User Architecture Specification

While the base configuration file and the data-width dependent configuration files
describe the base architecture properties, the user architecture configuration file
describes user specific settings. The tool-flow structure allows the user to assign
a unique architecture configuration file to each benchmark or allow reusing a
description of a (reconfigurable) architecture. As mentioned earlier, Blocks can
be instantiated with either a reconfigurable network (using switch-boxes) or with
fixed wiring. The latter produces an architecture instance that is essentially an
application specific VLIW processor with optimized data-path. Generally, this is
the first step in designing an architecture instance for Blocks, as this architecture
can later be mapped, as a virtual architecture, on a fully reconfigurable architecture
(with sufficient resources). An advantage of using the virtual architecture for initial
development is to reduce simulation time and to simplify debugging.

Specifying a Fixed Architecture

To configure a Blocks architecture instance with fixed wiring, all that the user
has to do is to add the desired function units to the 'architecture.xml' file and
connect the desired inputs and outputs of these function units together. Generally,
the way such a structure is built starts out with instantiating the required instruction
decoders. These instruction decoders are then connected to the output of an ABU to
receive a program counter. Listing 4.6 shows this for an example that implements
an architecture instance aimed at scalar binarization operations. Some of the
instantiated instruction decoders have been omitted for clarity.

```
 1  <functionalunits>
 2    <fu type="ID" name='id_abu'>
 3      <input index="0" source="abu.1"/>
 4    </fu>
 5
 6    <fu type="IU" name='imm'>
 7      <input index="0" source="abu.1"/>
 8    </fu>
 9
10    ...
11
12    <fu type="ID" name='id_alu_loop'>
13      <input index="0" source="abu.1"/>
14    </fu>
15
16    <fu type="ABU" name='abu' ID="id_abu" config="1">
17      <input index="0" source="alu_loop.1"/>
18    </fu>
19  <functionalunits>
```

Listing 4.6 Instruction specifications for the ALU

```
 1   <functionalunits>
 2
 3     ...
 4
 5     <fu type="LSU" name='lsu' ID="id_lsu">
 6        <input index="0" source="imm.0"/>
 7        <input index="1" source="alu.1"/>
 8     </fu>
 9
10     <fu type="ALU" name='alu' ID="id_alu" config="1">
11        <input index="0" source="imm.0"/>
12        <input index="1" source="lsu.0"/>
13     </fu>
14
15     <fu type="ALU" name='alu_loop' ID="id_alu_loop" config="1">
16        <input index="0" source="alu_loop.0"/>
17        <input index="1" source="alu_loop.1"/>
18        <input index="2" source="imm.0"/>
19     </fu>
20   <functionalunits>
```

Listing 4.7 Instruction specifications for the ALU

In this listing, it can be observed that all instruction decoders are connected to the output of the same ABU by having their only input connected to 'abu.1'. This is also the case for an immediate unit, named 'imm'. Additionally, an ABU is instantiated and configured to perform branching operations (instead of accumulate operations). This is done by setting its configuration bit to one. The ABU then produces a program counter that is used by the instruction fetchers and decoders. The ABU also takes an input of which the source is an output of an ALU. This ALU is used to compute the loop boundary and flags the ALU when the required number of iterations has been reached.

To finalize the construction of the architecture instance the function units themselves have to be added to the configuration. The instantiation thereof is very similar to those for the instruction decoders and ABU. Listing 4.7 shows the instantiation of the function units. The instantiation of the instruction decoders is omitted.

The data-width of the instantiated architecture can be chosen by selecting one of the data-width specific configuration files. Depending on which of the data-width specific configuration files is included the hardware will operate at a specific data-width.

Once complete it is possible to generate a graphic representation of the architecture; this aids in documentation and verification that the specified architecture is indeed the one that was intended by the designer. Figure 4.4 shows the generated graphical architecture representation for the scalar binarization example used in this section. The instruction decoders, shown in green, connect to the two ALUs (blue) and the LSU (red) to control their operation. Likewise, the ABU (pink) receives operations from its instruction decoders and generates a program counter (red arrows) that is used by all instruction decoders.

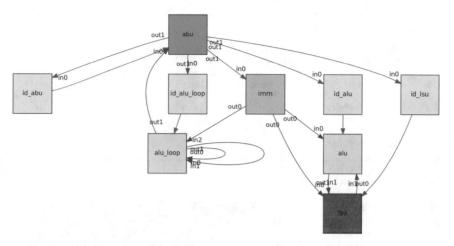

Fig. 4.4 Example of a generated graphical architecture representation for a Blocks architecture instance with fixed connections

```
1  <network>
2    <data horizontal='1' vertical='2'/>
3    <control horizontal='1' vertical='1' width='16'/>
4  </network>
```

Listing 4.8 Instruction specifications for the ALU

Specifying a Reconfigurable Architecture

Reconfigurable architectures are specified in a similar way to those with fixed connections. However, instead of specifying the connections between function units, properties of the interconnect can be specified. The interconnect is generated by the hardware generation tool described later in this chapter. The network configuration is specified for both the data and the control network and consists of the number of connections horizontally and vertically (between switch-boxes). For the data network, the width is inherited from the data-width dependent configuration file. In the example shown in Listing 4.8, the width of the control network is overridden; it shows how parameter overriding works within the Blocks configuration files.

Similar to the fixed architectures, function units are instantiated in the architecture configuration file. Listing 4.9 shows an example for an architecture that supports the scalar binarization example used in the previous section. The main difference, compared to the description for the fixed architecture, is that no connections are specified since these will be made at run-time using the switch-boxes. Instead, the instantiated function units have a parameter that determines their physical location within the generated physical architecture.

The specified hardware description can, just like the fixed architecture version, be viewed to check whether the intended design is achieved. This is performed using

```
 1   <functionalunits >
 2     <fu type="ID"    name="id_X0Y3" Xloc="0" Yloc="2"/>
 3     <fu type="ID"    name="id_X1Y3" Xloc="1" Yloc="2"/>
 4     <fu type="ID"    name="id_X2Y3" Xloc="2" Yloc="2"/>
 5     <fu type="ID"    name="id_X1Y1" Xloc="1" Yloc="0"/
 6     <fu type="IU"    name="imm_X0Y1" Xloc="0" Yloc="0"/>
 7     <fu type="ABU"   name="abu_X0Y2" Xloc="0" Yloc="1"/>
 8     <fu type="ALU"   name="alu_X1Y2" Xloc="1" Yloc="1"/>
 9     <fu type="ALU"   name="alu_X2Y2" Xloc="2" Yloc="1"/>
10     <fu type="LSU"   name="lsu_X2Y1" Xloc="2" Yloc="0"/>
11   </functionalunits >
```

Listing 4.9 Instruction specifications for the ALU

Fig. 4.5 Example of a generated graphical architecture representation for a reconfigurable Blocks architecture instance. It shows the data network

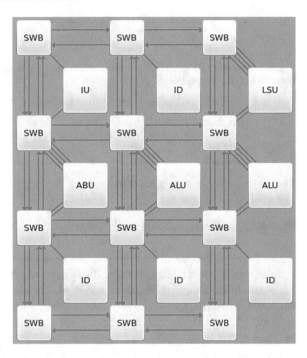

a graphical tool that can also perform automatic placement and routing. Figure 4.5 shows the reconfigurable fabric that allows mapping of the binarization application and architecture instance. As can be observed, the switch-boxes (marked with 'SWB') have one connection in each direction horizontally and two vertically as specified in the configuration file; these values can be changed by the user. Besides these connections, the switch-boxes connect to the function units. The function units are placed on the specified locations.

To run an application on the reconfigurable platform two things need to be specified: the placement, i.e. which function unit will be used for what purpose, and the routing over the switch-boxes. The routing is specified using a 'place-and-route' XML file located together with the benchmark application. This file can either be manually specified or be automatically generated by an automated routing tool. This content of this file will be explained in further detail in Sect. 4.5.

4.3 Hardware Generation

The Blocks tool-flow can generate the Hardware description language (HDL) for virtual and physical architectures. Some parts of the architecture are always the same and do not depend on the configuration. Examples are the boot-loader and the internal structure of function units. These building blocks can, therefore, be simply instantiated and used in the architecture. Any parameters that need to be configured, such as the data-width, are inherited from the higher hierarchy levels. Other parts of the architecture are generated based on the configuration files supplied by the user, like the compute fabric structure (which FUs are located where), the interconnect network, and the switch-boxes themselves. All templates have special tags that are replaced by the hardware generation tool.

The hardware generation tool loads the hardware description as specified by the user, including all base configuration files. The hardware description is then checked against a set of design rules to prevent the user instantiating a non-functional design. If any of these rules is violated, the tool will respond with an error message telling the user where the problem can be found and what the cause of the error is. Examples of such design rules, but not limited to, are: check if all referenced instruction decoders exist; check that all referenced input sources exist in the design; and check that at least one of the outputs of a function unit is used (otherwise the unit is 'dangling').

Once the design passes the DRC check, the tool begins searching for special tags in the templates. The tags are always in the form $<< TAG >>$. Depending on the name of the tag, the hardware instantiation tool replaces the tag with a value or an entire HDL structure. A value can be, for example, the data-width. An example of an entire structure is the instantiation of a function unit.

4.3.1 The Compute Template

The most important template in Blocks is the 'compute' template. This file instantiates the main HDL file that contains all user specified function unit instances as well as the interconnect. Since the whole compute section depends on the user configuration, most of the HDL content for this template is generated by the hardware instantiation tool. Figure 4.6 shows an example of what is instantiated inside the compute template. For clarity, the interconnect is not drawn. In this example, the instantiated instruction decoders and function units are shown. To allow communication to the higher hierarchy levels within the Blocks structure, the

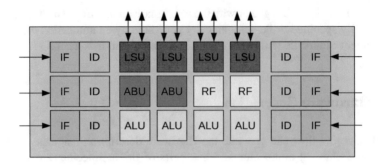

Fig. 4.6 Example of components instantiated in the compute template

compute template is also used to instantiate memory interfaces for the LSUs (both local and global) as well as the memory interfaces for the instruction decoders. Generation of the data and control networks is performed by the hardware generator and both can be either a fixed or a reconfigurable network.

In the first case, the connections described in the architecture configuration file are instantiated using directly wired buses in HDL. Hence, they cannot be modified at run-time. This method is useful for debugging and to test functionality of the intended architecture on FPGAs. The reconfigurable network is generated by building a network out of switch-boxes. These switch-boxes are generated for each location in the network as their functionality with regard to the number and location of input and output ports depends on their location. For example, a switch-box on the top left of the network does not require connections on the left and top of the switch-box. Similarly, switch-boxes that are not located next to a function unit will not require function unit connections. The topology of the network can be specified in the configuration file and is generated by a module within the hardware generator. Currently, only a full mesh network is supported. However, the network generation module is designed such that it can easily be replaced or extended to support other network configuration schemes. Besides the HDL describing the functionality of the switch-box, a file is generated that describes the routing paths available in each switch-box. These files are used in the bit-stream configuration tools to configure the desired paths on the fabric.

Once the function units and network have been instantiated a scan-chain is constructed to allow configuration of the units and switch-boxes. The scan-chain passes through each FU and switch-box that requires configuration and consists of one data wire and an enable wire. With the enable wire high the data is clocked in and shifted over the data wire that passes between all nodes in the chain. The scan-chain is controlled by the boot-loader. The order of the nodes in the scan-chain is written to a file that is used by the bit-stream generation tool. The boot-loader is contained in the next level up in the hierarchy, called the 'compute wrapper'. This structure of this template is shown in Fig. 4.7.

The compute wrapper contains, besides the boot-loader and the memory arbiter, a module called the 'state controller'. The state controller implements a secondary

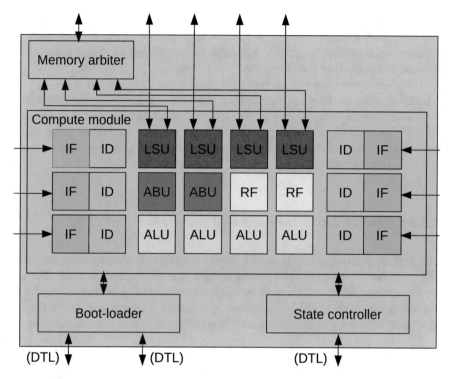

Fig. 4.7 The compute-wrapper template instantiates the boot-loader and, if enabled, a state controller that can save the state of all registers at a certain point during execution

scan-chain that allows reading all registers inside the compute region. This can be used for in-system debugging as well as context switching. The memory arbiter allows all LSUs to access the global memory by managing read and write requests. The arbiter requires a connection to a higher hierarchy level for its connection to the actual global memory. Similarly, the memory interfaces for the local memories and the instruction memories are made available to a higher hierarchy level. This hierarchy level allows synthesis of the parts of Blocks that perform computation without requiring synthesis of the memory modules, thus allowing for easy control of a test-bench on this level, if required.

4.3.2 The Core Template

The core template is a layer around the Blocks compute area and a memory module; it connects all internal wiring between the compute area and the local memories, and the instruction memories. The core level in the hierarchy is the level that would be used by a designer wishing to integrate a Blocks fabric instance in a System-on-chip (SoC). At this level, the configuration interface, the boot-loader interface, and the global memory interface are available as buses that can be connected to, for example, a network-on-chip or AXI interconnect. Figure 4.8 shows the memory module within the core level hierarchy.

The memory module can instantiate various types of memories depending on what is specified in the configuration files. For example, an FPGA implementation uses other memory primitives than an ASIC implementation or Register transfer level (RTL) simulation. The memories inside the memory module do not perform handshaking and can, therefore, not be stalled; this is to improve performance and reduce system complexity.

4.3.3   The Top Level Template and Test-Bench

The highest hierarchical level in the Blocks templates is the top level template. This template is required when Blocks is used as a stand-alone platform. The top level contains a secondary memory arbiter that can be used by an external device to preload the contents of the global memory with data. Through this arbiter Blocks gains access to the global memory and, if configured, to one or more peripherals.

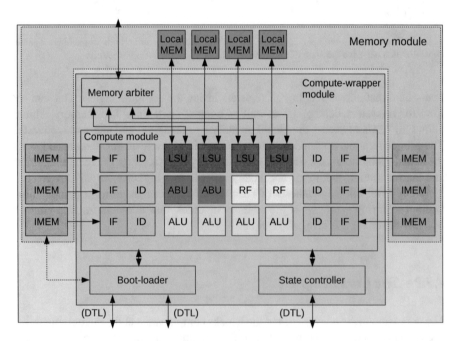

Fig. 4.8 Example memory module that contains local memories and instruction memories. The core level hierarchy connects the memory module to the compute module of the Blocks architecture

These peripherals can be instantiated based on descriptions in the configuration files, including external inputs and outputs.

Around the top level a test-bench is generated that provides simulation for external memories for application storage as well as global memory data preloading. The test-bench then controls application execution and allows read-back of the computed results that are written in the global memory during execution. Figure 4.9 shows an example top level module with the attached test-bench.

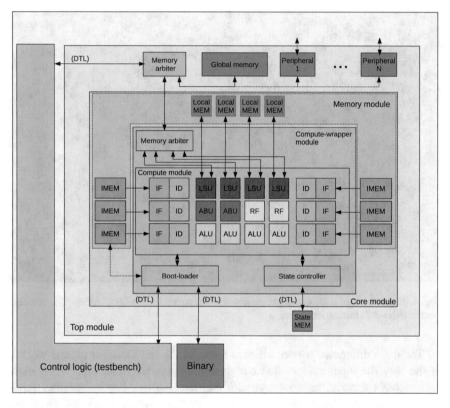

Fig. 4.9 Example top level module for a stand-alone Blocks architecture instance, with attached test-bench for application simulation

4.4 Programming Blocks

Programming Blocks is in many ways similar to programming a VLIW processor in assembly. Like a VLIW, there are multiple parallel issue slots that operate in lock-step. Each of these issue slots performs its own operations. In the Blocks

assembly language, called Parallel assembly (PASM), each issue slot has its own
column. These columns contain the operations that are specific to an issue slot.
There is no difference if this issue slot controls a scalar or vector operation as one
instruction decoder may simply be connected to more than one function unit. The
(reconfigurable) network dictates where the source operands come from and where
the results will go to. Figure 4.10 shows an example application written in PASM,
and each column corresponds to an instruction decoder.

	id_abu	id_alu_loop	id_lsu
1			
2			
3	.text	.text	.text
4			
5	nop	nop	nop
6	nop	nop	nop
7	nop	nop	srm r9, in0
8	nop	nop	srm r13, in0
9	nop	pass out0, in2	nop
10	nop	nop	srm r2, in0
11	nop	nop	srm r5, in0
12	nop	pass out1, in2	nop
13	nop	nop	nop
14	nop	nop	lgi BYTE, out0
15	nop	sub out1, in0, in1	lgi BYTE, out0
16	nop	sub out1, in0, in1	lgi sgi BYTE, out0, in1
17	bcri -3, in0	sub out1, in0, in1	lgi sgi BYTE, out0, in1
18	nop	sub out1, in0, in1	lgi sgi BYTE, out0, in1
19	nop	sub out1, in0, in1	lgi sgi BYTE, out0, in1
20	nop	nop	sgi BYTE, in1
21	nop	nop	sgi BYTE, in1
22	jai 0	nop	nop

Fig. 4.10 Example of a Blocks application written in parallel assembly (PASM). The columns
control individual instruction decoders

The main difference between Blocks applications and those for typical VLIWs
is the way the input sources and output destinations are described. While most
architectures specify input sources as a register location, Blocks specifies input
sources as an input port of the function unit. The reconfigurable network, therefore,
defines what the connection to a certain input port actually means. Similarly, the
output describes which output port is used to make the result available on the
network. An application for Blocks, therefore, consists of two parts: an architecture
description and the instructions that are going to be executed on this architecture
instance.

4.5 Configuration and Binary Generation

After the hardware has been generated by the tool-flow, it is still required to configure the design at run-time. This has to be done at two levels: a bit-file is loaded that configures the function units and the switch-boxes in the reconfigurable fabric, followed by moving the instructions found in the executable binary into the local instruction memories. The latter is required to provide sufficient bandwidth to load instructions in parallel as the interface to external memories cannot supply this bandwidth. The data for both of these steps are located in a single binary file that is generated by the Blocks tool-flow, as shown in Fig. 4.2. Various steps of this process are performed by different tools, providing separation in functionality. These tools are described in the following subsections.

4.5.1 Assembler

The assembler[1] translates the human readable assembly code into machine instructions. It takes the user program as an input, together with the hardware description. The hardware description is required as it describes the translation of human readable mnemonics into machine code. In essence, this provides a look-up table that is used to perform the required translation.

Most instructions contain arguments. These arguments should be of a certain type, for example an input or output, and within a certain range. The assembler checks the arguments and returns an error if the proper set of arguments is not provided for a specific instruction. The required arguments and their properties are specified in the hardware configuration file as well. This means that if new instructions, or even function units, are added to the architecture template, it is not required to modify the assembler.

The assembler supports replacing labels that are used in jump or branch instructions with the actual instruction address. For each column, and therefore for each instruction decoder, the assembler generates a file with the encoded machine instructions. These files can be directly used for unit testing of function units or used by the binary generator tool to produce the final executable.

4.5.2 Bit-File Generator

Some function units require setting configuration bits before use. For example, the ALU has a configuration bit that determines if an output is buffered or unbuffered.

---

[1] This tool was designed together with Luc Waeijen.

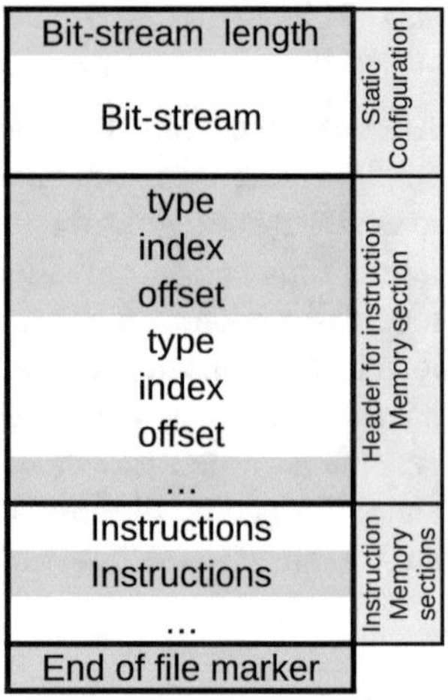

Fig. 4.11 Structure of the
executable binaries as
generated by the binary
generator. The bit-stream
specifies static configuration

Bit-stream length	Static Configuration
Bit-stream	
type	Header for instruction Memory section
index	
offset	
type	
index	
offset	
...	
Instructions	Instruction Memory sections
Instructions	
...	
End of file marker	

Similarly, instruction decoders have to be configured for what type of function unit
they will be controlling. Part of the values for these configuration bits is specified
in the architecture configuration; others are determined by the hardware generation
tool. A Blocks architecture with fixed connections can be automatically analysed to
infer some of the required configuration values. For example, an instruction decoder
that is connected to an ALU will have to be able to control this type of function unit.
For reconfigurable architecture instances the configuration values are contained in a
place-and-route file. This file contains the switch-box configurations to implement
the desired paths on the network as well as the function unit configurations. This
file can be generated with an automatic placement and routing tool, described in
Sect. 4.5.4.

The bit-file generator loads the required architecture configuration files as
well as a description provided by the hardware generator that specifies the order
of configurable items in the configuration scan-chain. Based on this order the
individual unit configurations are concatenated into a single bit-stream. The bit-
stream is passed to a binary generator that merges the generated application code
with the bit-stream.

4.5.3 Binary Generator

The binary generator gathers all generated machine code files as well as the bit-
stream and merges these into a single executable file that can be processed by
the boot-loader of Blocks. The binary consists of several sections that provide
information to the boot-loader where specific instruction regions need to be placed.
The structure of the binary is shown in Fig. 4.11.

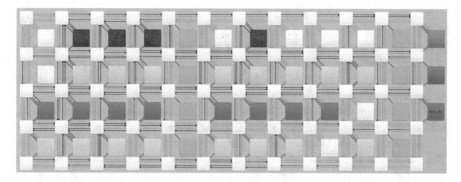

Fig. 4.12 Example placement and route of a vector binarization algorithm with eight vector lanes using the automated placement and routing tool

The binary starts with a field that indicates the length of the bit-stream section enclosed in the binary. This tells the boot-loader how many memory words will have to be read to configure the device. The bit-stream is zero padded at the front of the stream such that the length of the bit-stream is always a multiple of the memory width, simplifying loading. Once the bit-stream has been loaded, and the hardware configured, an index specifies, for each instruction decoder, where its instruction segment can be found. The boot-loader iterates through this list and loads the corresponding instruction segments into the dedicated instruction memories. A special instruction segment type indicates the end of the list after which boot-loading is complete.

4.5.4   Automatic Placement and Routing Tool

Manually specifying the placement and routing configuration is both time consuming and error prone. For this reason, an automatic placement and routing tool has been developed.[2]

The routing tool takes the reconfigurable architecture description as a source file for the available function units, the switch-boxes, and the connections between these switch-boxes. The second input is an architecture description of a CGRA with fixed connections and function units. These function units are then placed on the fabric, and the connections, as described for the fixed architecture, are routed over the reconfigurable network. Figure 4.12 shows an example of a placed-and-routed CGRA instance. The gray units and connections are resources that are not used in this CGRA configuration.

---

[2]The placement and routing tool was developed by Coen Beurskens as part of his internship in the Electronic Systems Group at Eindhoven University of Technology.

Simulated annealing is used for placement of the function units on the reconfig-urable fabric. In most cases there are multiple options of how function units can be placed. The actual placement may have a large influence on the routing step that is performed after placement. The tool performs very well, usually matching and sometimes beating handmade configurations with respect to the total path length and average path length.

For placement with simulated annealing it is required to start with a first, valid, solution of placement into the fabric. This initial placement is random but places function units only on locations where that function unit type can actually be placed. The placement is then modified by the simulated annealing algorithm based on a cost function and a decreasing 'temperature'. The cost function is designed to minimize the total routing length and minimize network congestion. In simulated annealing the temperature slowly decreases; once the temperature multiplied with the result of the cost function comes below a threshold, the algorithm is stopped.

Routing is based on the PathFinder algorithm [2]. This algorithm initially routes all nets using the shortest path algorithm and may route multiple nets onto the same physical channel. An iterative process of ripping up connections and re-routing them over free resources then optimizes the routing and makes sure each physical channel is only used by a single net. Which net stays and which net will be rerouted depend on a penalty function. This function depends on the occupation of the channel and the gain a net receives from being routed on that channel. Optimizations such as wavefront routing allow the reuse of existing channels for broadcasting to multiple destinations, thus preventing a separate route for every destination.

4.6   Conclusions

Every computer architecture requires a supporting tool-flow to allow designers to use it. The tool-flow for blocks uses a physical architecture description to specify the resources that are available and how these can be connected. On top of this, a virtual architecture specifies the function units that are used and the connections between them. The physical architecture description is used to generate synthesizable HDL that can be processed into a chip layout via ASIC tools. The virtual architecture is used to generate a configuration bit-stream for the physical architecture as well as an executable binary. It is also possible to use a virtual architecture description to generate an FPGA implementation for fast prototyping.

The tool-chain consists of several template based tools. These tools generally take an XML file as an input and process these into configurations or HDL descriptions. Virtually everything in the hardware generation is specified in the input XML files, making the tool-flow very flexible and allowing easy extensions to the Blocks architecture framework.

Blocks is programmed using a special assembly language called PASM (parallel assembly). This language uses multiple columns to control the individual IFIDs present in the design. Every IFID instruction on a line, if connected to the same

ABU, is operating in lock-step. The PASM file is translated into an executable binary by the tool-flow. Another tool generates a bit-stream that describes the network and function unit configuration. The resulting bit-stream is appended to the executable binary to allow configuration and program execution on Blocks.

References

1. S. Goossens, et al., The CompSOC design flow for virtual execution platforms, in *Proceedings of the 10th FPGAworld Conference. FPGAworld'13* (ACM, Stockholm, Sweden, 2013), pp. 7:1–7:6. ISBN: 978-1-4503-2496-0. https://doi.org/10.1145/2513683.2513690. https://doi.acm.org/10.1145/2513683.2513690
2. L. McMurchie, C. Ebeling, *PathFinder: A negotiation-based performance-driven router for FPGAs*, in *Third International ACM Symposium on Field-Programmable Gate Arrays*, Feb. 1995, pp. 111–117. https://doi.org/10.1109/FPGA.1995.242049

Chapter 5
Energy, Area, and Performance Evaluation

Previous chapters describe the concept of the Blocks architecture framework. However, to know how well Blocks performs with respect to other architectures, both reconfigurable and fixed, is crucial. Unfortunately, comparing computer architectures based on published results is hard. There are various reasons to this: benchmarks are not the same, memories and their interfaces are often not taken into account, architectures are only functionally simulated, and energy numbers are either not mentioned, or it is not clear what is included in these numbers and what is left out. Besides this, there are often differences in the chip manufacturing technology used. For example, one design may use a 40 nm process, while another uses a 28 nm process. Although it is possible to scale these numbers, it leads to inaccuracies in the comparisons. Since many other architectures do not provide full RTL, it is not really possible to synthesize and place and route these for the same technology as used for Blocks to obtain comparable numbers. This is beside the significant amount of work that is typically required to reach this stage.

In order to provide a proper evaluation of the main innovation of Blocks, separating the control from the data using two separate FPGA-like networks, the reconfigurable reference architectures are based on the Blocks framework. The reference architectures use special versions of the Blocks function units that include a local instruction decoder and instruction memory. By doing so, all other variables can be eliminated and a proper energy comparison can be made. A similar method is used to generate the fixed SIMD and VLIW processor architectures. These use fixed-connection architecture instances generated by the Blocks framework to implement the required processors. Therefore, there is no difference between the instruction set architectures of the reference architectures and Blocks, with exception of an ARM Cortex-M0 microcontroller that is used as a general purpose microprocessor reference architecture. All architectures are evaluated for performance, energy efficiency, area, and area efficiency. These metrics are described in more detail in Chap. 1.

M. Wijtvliet et al., *Blocks, Towards Energy-efficient, Coarse-grained Reconfigurable Architectures*, https://doi.org/10.1007/978-3-030-79774-4_5

Section 5.1 of this chapter introduces the Blocks instances and reference architectures used to perform the evaluation. Section 5.2 details the benchmark kernels that are executed on these architectures. Section 5.3 presents the results and an evaluation thereof. Finally, Sect. 5.4 concludes this chapter.

5.1   Architectures

In order to obtain accurate results, all architectures are placed and routed using commercial ASIC design tools targeting a commercial 40 nm technology library. Power and energy results are based on real toggling rates on the full layout, and these are obtained by simulating the generated net-list for all applications. Four reference architectures are used for comparison: a reference CGRA, an 8-lane SIMD, an 8-issue VLIW, and an ARM Cortex-M0 microprocessor. Additionally, application specific processors (ASPs) are instantiated per benchmark application. The SIMD, VLIW, and ASP architectures are all realized with the Blocks framework but use fixed connections instead of a reconfigurable fabric.

All architectures contain a two-ported (one read and one write port) global data memory of 32 kilobyte, sufficient to hold the input and output data for the evaluated benchmarks. The ARM-M0 uses a single ported memory due to AHB bus limitations. All LSUs, with exception of the LSU embedded in the ARM Cortex-M0, contain a 1 kilobyte two-ported local memory. The local memories are private to an LSU and, therefore, does not need arbitration. Its size is chosen such that it allows some local data reuse without allowing the whole input dataset to be made available locally.

The overall goal is to ensure fair comparisons as much as possible. For this reason, the compute area is designed to be as similar as possible between Blocks, the traditional CGRA, VLIW, and SIMD. The compute area is considered to be the area occupied by the architecture excluding the area for memories and switch-boxes. Any remaining area differences are corrected for by using the performance per area metric.

Table 5.1 summarizes the architecture properties for all architectures. All architectures were synthesized with Cadence RTL compiler (14.11.000) and the layout was achieved using Cadence Innovus (16.13.000).

5.1.1   Blocks Architecture Instance

The instantiated (reconfigurable) Blocks architecture is reused for all kernels (described in Sect. 5.2) and contains the superset (union) of function units in the ASPs. It includes 9 LSUs, 17 ALUs, 9 multipliers, 2 immediate units, 1 RF, and 1 ABU, as shown in Fig. 5.1. When an FU is not in use for a kernel, the inputs are tied to a fixed value, by configuration, to reduce power. The compute resources are very similar to the 8-lane reference SIMD with control processor, but the RFs

Table 5.1 Summary of architecture properties

Architecture	GM [KB]	LM [KB]	IM [KB]	#ALU	#MUL	#LSU	#RF	#IM	Compute area [mm^2]
VLIW	32	4	14.8	8	8	4	8	4	0.18
SIMD	32	9	4.4	9	9	9	9	1	0.23
ARM M0	32	0	32	1	0	1	1	1	0.02
Traditional CGRA	32	9	15.5	17[a]	9	9	1	2	0.26
Blocks	32	9	5	17[a]	9	9	1	2	0.26

[a]Eight of these ALUs are only used as 2-register data buffer

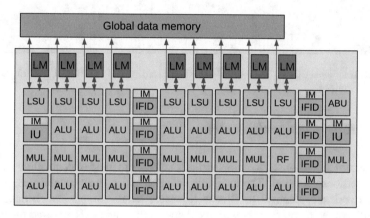

Fig. 5.1 Blocks architecture instance used for the benchmark kernels. For clarity the networks (described in Chap. 3) are omitted

are replaced by ALUs, which are used as two-element register files in the 'FIR', 'IIR', and '2D convolution' benchmarks, and are unused in other kernels. These benchmarks do not require many registers due to spatial layout; in this case, ALUs are cheaper in area and power. When performance per area is used as a metric, this still results in a fair comparison, allowing direct performance comparison. The Blocks instance contains 8 instruction decoders to control groups of these function units. The two networks (control and data) have different widths and a number of connections per switch-box. The data network in all Blocks instances is 32-bit wide (but can be configured at design time to be 8-, 16-, 32-, or 64-bit wide). The control network has a fixed width of 16 bit and transports the decoded instructions. The instantiated data network allows 3 vertical and 2 horizontal connections (on each side) per switch-box, and the control network allows 1 vertical and 2 horizontal connections per switch-box. These connections are sufficient for implementation of the kernels in the benchmark set. The interconnect pattern is a full mesh. Both the number of connections and interconnect pattern are design parameters.

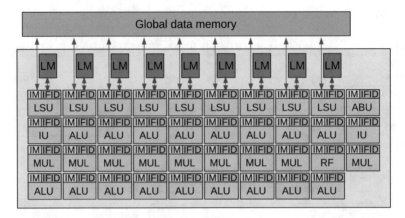

Fig. 5.2 Traditional CGRA. The instruction memories are marked with 'IM'. For clarity the data-network is omitted

5.1.2 Reference Architectures

A reference CGRA is used to compare the traditional method of controlling a CGRA with the Blocks method. The available function units, their capabilities, and the data network are identical to Blocks to ensure a fair comparison. Blocks-based application specific processors with fixed connections and FUs are used to evaluate the overhead of Blocks. The VLIW and SIMD processors are used to compare the performance and energy efficiency of Blocks with processor types typically used for DSP applications. The goal is to evaluate whether Blocks can achieve a similar performance per area, and energy efficiency, as the best suited general purpose processor for each benchmark kernel. The SIMD and VLIW processors feature nearest neighbour connections between lanes/slots and provide bypassing capabilities between pipeline stages, ensuring a fair comparison with Blocks. The ARM Cortex-M0 is used to compare performance per area and energy with a general purpose low-power microprocessor.

Traditional CGRA

The traditional CGRA is a special version of Blocks without separation of the control and data paths. It is referenced throughout this work as 'traditional'. It contains the same specialized function units as Blocks, but with local instruction memories for *each* unit, as shown in Fig. 5.2. Since each function unit performs local instruction decoding, the control network is no longer required and is removed. This CGRA is used as a reference since comparisons with existing CGRAs are often inaccurate. For most existing CGRAs there are no power or energy numbers reported. And, if they are they are (see chap. 2) often for different technology nodes. Furthermore, what is included in the energy numbers is often unclear (e.g., if it includes memory accesses). In our opinion, using a Blocks-based reference CGRA provides the fairest comparison to show the benefits of separating the control-path and the data-path.

Application Specific Blocks Instances (Blocks-ASP)

The application specific Blocks instances contain fixed wiring between function units instead of the reconfigurable data and control network. For each application, the unused function units are removed. This results in a dedicated processor that contains only the hardware (connections, function units, memories) that is required for executing a specific kernel, called Blocks-ASP. The architecture and schedule of these Blocks-ASP processors are mapped onto both Blocks-ASP and Blocks (with a reconfigurable network) such that the Blocks reconfiguration overhead can be determined. The structure of the implemented processors is as described in Sect. 5.2.

ARM Cortex-M0

The ARM Cortex-M0 was chosen for its popularity in commercial applications and its reputation as a very low-power architecture. It uses a Cortex-M0 logic core combined with an instruction and data memory, as shown in Fig. 5.3. Both memories are accessed by the single Cortex-M0 memory port on the processor. Normally, this port interfaces to a single memory, but for energy evaluation the memory is split into two separate memory blocks. These memory blocks are mapped by address range. One memory contains all data and the other all instructions. This allows comparison of instruction memory and data memory access energy with respect to the other architectures. The largest binary for this architecture is approximately 12 kilobyte, but since the same memory is also used as Random access memory (RAM), 32 kilobyte memory is required. The instruction memory is single ported.

In order to evaluate if a more powerful processor in the ARM Cortex series would make any difference, the cycle accurate processor simulator OVPsim was used to predict performance for the Cortex-M4f and the Cortex-A9. The power numbers specified on the website of the manufacturer are used to estimate any possible differences in energy with respect to the Cortex-M0. This is described in more detail in Sect. 5.3.2.

Fig. 5.3 ARM Cortex-M0 reference architecture

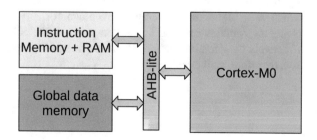

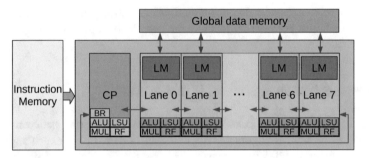

Fig. 5.4 8-Lane SIMD reference architecture

8-Lane SIMD with Control Processor

The SIMD reference processor is based on the architecture presented in [1]. It contains a control processor and 8 lanes with bypass support between ALU, multiplier, and LSU. The control processor controls branches and can broadcast values to the lanes in the vector processor, as shown in Fig. 5.4. All lanes contain a register file, ALU, multiplier, and LSU. The lanes and control processor feature a neighbourhood network, allowing results to be communicated between lanes. The instruction memory of the SIMD holds up to 256 instructions, the next power of 2 needed to fit the largest kernel. Only one of the function units inside each lane (ALU, multiplier, LSU) can be active per clock cycle; this is typical for SIMD processors. The neighbourhood network gives it a small advantage over regular SIMD processors by allowing results to be directly passed to a neighbouring lane. This increases both performance and energy efficiency.

8-Issue Slot VLIW

The VLIW reference architecture has eight issue slots with private register files, neighbourhood communication, and bypass support (Fig. 5.5). If a VLIW with a shared RF would have been used the power, results would have been worse due to the many required register ports. One of the issue slots supports branching. The instruction memory holds up to 256 instructions, although these instructions are significantly wider than for the SIMD. Half of the issue slots contain an LSU, which is a higher than average percentage for a VLIW. Although the VLIW processor contains half the LSUs compared to the SIMD, it can use them more effectively by spreading data fetches over time, which prevents memory congestion and, therefore, stall cycles.

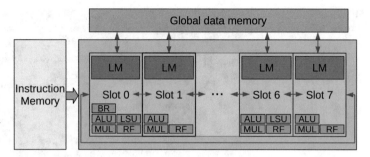

Fig. 5.5 8-Issue slot VLIW reference architecture

5.1.3   Synthesis and Place and Route

Synthesis and place and route of the reference architectures are performed using the Cadence ASIC tools on a commercial 40 nm low-power library. To obtain accurate results, all memories are implemented using commercial 40 nm memory macros that can be taped out. The resulting net-list also includes the clock-tree and I/O buffers. For the reconfigurable versions of Blocks, timing constraints had to be specified in a somewhat convoluted way. The switch-boxes are circuit switched and therefore combinational, and due to this, the ASIC tool-flow finds timing loops and, therefore, cannot guarantee any timing. Since it is crucial to predict performance, timing has to be constrained. The solution was to add buffers to the outputs of the switch-boxes and cut the timing arcs in the corresponding cells after elaboration. These buffers can then be used as timing start and end points. It now becomes possible to specify timing constraints between switch-boxes. A similar technique is used to constrain the function unit delays. Although the function units are registered at the end of a path, the path from a switch-box output to the register also has to be specified. This constrains the propagation time through the function units and switch-boxes, but it does not determine the maximum frequency of an instantiated architecture, as this depends on the actual placement and routing on the Blocks fabric. The advantage is that for each sub-path (switch-box to switch-box, switch-box to FU, etc.) of the design, the maximum propagation time is known. This information can be used by the Blocks place-and-route tools for optimization and to determine the maximum clock frequency a priori, as described before.

Performance and power results are obtained by simulating the placed-and-routed net-list at the typical corners including delay annotation and capacitances. This has the advantage that functionality can be verified after the whole design flow, as well as obtaining an accurate activity file (TCF). The net-list combined with the activity file is used by the Cadence flow to perform power analysis. Since this is performed on a post place-and-route design, wire and cell capacitances are included. The same technology library and synthesis settings are used for all architectures and all benchmarks.

Fig. 5.6 Post
place-and-route layout of
Blocks

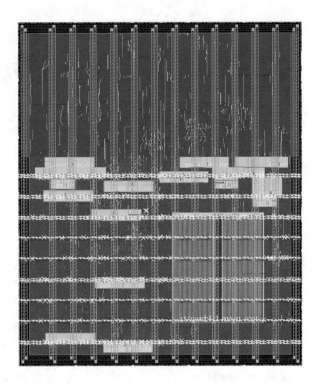

5.1.4   Critical Path Analysis

In this work, all designs are synthesized for 100 MHz in order to be able to make a fair comparison. At this frequency, there are not many effects from the synthesis tools selecting larger gates to reach the desired clock frequency. In practice, Blocks has been successfully placed and routed at 300 MHz (worst case corner at 125 °C and 0.99 V). The reference architectures can be synthesized up to 300 Mhz, 470 MHz, 300 MHz, and 450 MHz for the traditional CGRA, VLIW, SIMD, and ARM Cortex-M0, respectively. For Blocks, VLIW, and SIMD, the memory arbitration unit eventually becomes the critical path of the design. The maximum frequency is higher for the VLIW as it only contains four LSUs, while the SIMD contains nine. Each switch-box introduces a delay of 0.2 ns, and the worst case function unit, the multiplier, introduces a delay of approximately 2 ns. This means that even when the multiplier is used, it is still possible to route the signal over six switch-boxes before the maximum frequency has to be reduced. A higher achievable frequency allows voltage scaling and further energy reduction. However, Sect. 5.3 shows that Blocks obtains a significantly higher performance at the same frequency compared to the reference architectures. Therefore, Blocks can achieve similar performance while still scaling to lower voltages. Figure 5.6 shows the post place-and-route layout of Blocks. The yellow (LM and IM) and red (GM) blocks are memories, and the remainder is logic.

Table 5.2 Overview of benchmark kernels

Benchmark	Description	Type	Data size
Binarization	Thresholding on greyscale image	Scalar to scalar	128*64 pixels (8 bit)
Erosion	Noise removal by AND-ing neighbouring pixels	3*3 window to scalar	128*64 pixels (8 bit)
Projection	Sums each horizontal and vertical row/column	Vector to scalar	128*64 pixels (8 bit)
FIR	8-tap low-pass FIR filter on input signal	8*1 convolution	2200 samples (32-bit)
IIR	Third-order low-pass filter on input signal	2*1 + 2*1 convolution	2200 samples (32-bit)
FFT	8-point complex FFT on audio signal	8*1 to 8*1 vector (complex)	2200 samples (32-bit)
2D convolution	Gaussian blur on image	3*3 window to scalar	128*64 pixels (8 bit)
FFoS	Industrial vision application	Image to 2 8*1 vectors	128*64 pixels (8 bit)

5.2   Benchmark Kernels

Blocks aims at efficient execution of digital signal processing applications. To evaluate Blocks, eight benchmark kernels from the computer vision and signal processing domains are used that include both 1D and 2D operations. The type of parallelism available (data or instruction level) varies per kernel. The benchmark set is selected such that some kernels perform well on SIMD processors and others on VLIW processors. Table 5.2 provides an overview of the benchmarks used for evaluation. All benchmarks are optimized per architecture. Table 5.3 gives an overview of the Blocks resources that are used for each kernel implementation.

The kernels for all reference architectures, except the ARM Cortex-M0, are written in assembly language. The benchmarks for the M0 are written in C and compiled with the '-O3' flag using GCC, version 4.9.3. For all benchmarks and architectures, the initial data is assumed to be in the global data memory, and the local memories are considered to be uninitialized.

The remainder of this section describes a dedicated virtual Blocks architecture for each benchmark kernel. The architectures are designed such that they fit on a reconfigurable Blocks architecture with a function unit distribution similar to an eight-lane SIMD with control processor and an eight-issue slot VLIW.

Binarization The binarization kernel thresholds input values, typically from an image, to produce a 'one' if the pixel value is above the threshold or a 'zero' if it is below the threshold. This kernel is easy to vectorize since all input and output values are independent. When vectorized for a width of eight elements, this kernel requires eight LSUs for data loading and storage. Eight ALUs are used for processing and an additional ninth is used for loop boundary calculations, as shown in Fig. 5.7.

Table 5.3 Overview of resources occupied by the benchmark kernels

Benchmark	ABU	ALU	LSU	MUL	RF	IU	ID
Binarization	1	9	8	0	0	1	4
Erosion	1	9	8	0	1	1	5
Projection	1	9	9	0	1	1	6
FIR	1	16	1	8	0	1	6
IIR	1	10	1	8	0	1	7
FFT	1	13	8	8	0	2	8
2D convolution	1	16	4	9	0	1	8
FFoS	1	9	9	0	1	1	6

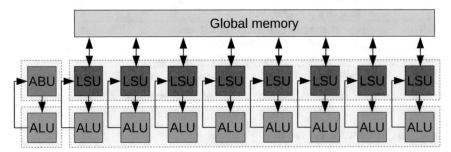

Fig. 5.7 Schematic overview of the implemented processing structure for the binarization kernel

An ABU performs program flow control and controls four instruction decoders to form a four-issue VLIW architecture with vector lanes. This allows parallel control of the ABU and an ALU for loop bound calculation (for program flow control) and a vector processor of ALUs and LSUs for computation. The regions shown around one or more function units indicate that these are under control of the same instruction decoder. Some paths, e.g., for setting up the addresses of the LSUs, are omitted for clarity.

Erosion Erosion is a two-dimensional operator sliding over the image in a way similar to convolution. However, the operation only performs a wide logic 'and' operation on neighbouring pixels. This is a commonly used denoising technique in image processing. It is simpler than convolution since there are no coefficients and the inputs and outputs are binary. The implemented algorithm computes the results in a single pass over the image. Horizontal sub-results are computed and stored as intermediate results into the local memories as a line buffer. When sufficient initial rows have been processed, the final results for each pixel can be computed and stored in the memory. The architecture for this implementation requires nine ALUs, eight for processing and one for loop boundary calculations. Furthermore, eight LSUs are used for data loading and storing, see Fig. 5.8. A register file is used for intermediate values and loop counter variables. The ABU therefore controls five instruction decoders.

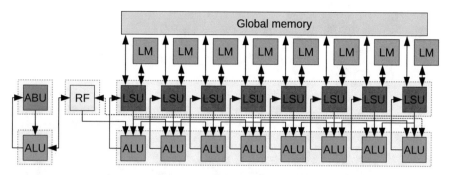

Fig. 5.8 Schematic overview of the implemented processing structure for the erosion kernel

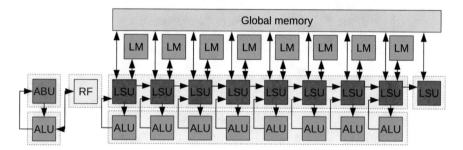

Fig. 5.9 Schematic overview of the implemented processing structure for the projection kernel

Projection Projection involves summing pixel values of an image in both horizontal (row) and vertical (column) directions. This results in two vectors that can be used for peak detection. The implementation of this algorithm computes both the horizontal and vertical summations in a single pass over the image. The architecture designed for this requires nine ALUs, eight for computation and one for loop boundary calculations. Nine LSUs are used, eight for data loading of the input vectors and a separately controlled scalar LSU for storing results for each completed row. Intermediate results for the vertical summations are kept in local memories and stored in the last pass to global memory. Since more bookkeeping has to be performed for the program flow, a register file is required. Figure 5.9 shows a schematic representation of this implementation. In total, the ABU in this design controls six instruction decoders.

FIR The Finite impulse response (FIR) is a very common digital filter. It consists of a single delay line that is used to feed multiplications with time-delayed data. The results of all multiplications are summed together to produce the output value. The implementation of this algorithm uses eight ALUs as a delay line for eight multipliers that compute the results for each filter tap. A reduction tree, controlled by a single instruction decoder, is used to compute the final FIR filter result. All steps are performed in a pipelined way, such that there is a throughput of one element per

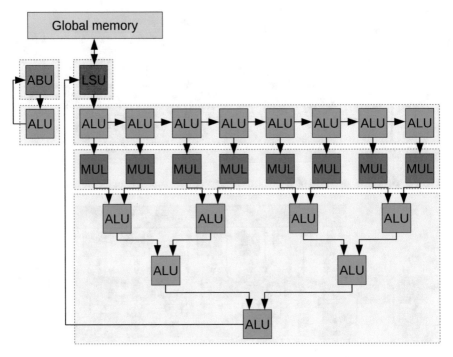

Fig. 5.10 Schematic overview of the implemented processing structure for the FIR kernel

(non-stalled) program cycle. Memory operations are performed by a single LSU, as can be seen in Fig. 5.10.

IIR An Infinite impulse response (IIR) filter is another commonly used digital filter. The main difference with the FIR filter is that (intermediate) results can be fed back into the filter. The implemented IIR filter has two filter stages, one of order one and another of order two. The filter is implemented in a spatial layout with as much SIMD control of these two stages as possible, see Fig. 5.11. In total, nine ALUs are used for summation of (intermediate) results. Eight multipliers are used for coefficient multiplication and a single LSU is used for memory operations.

2D convolution 2D convolution is often used for image filtering. Various coefficient windows can be used for blurring images, detecting edges, sharpening, etc. In all cases, convolution is a 2D window moving over the image and producing a single output element. The output pixels are independent, but the input data is reused between convolution steps. This leads to opportunities for data reuse, thereby avoiding accesses to global memory where possible. The algorithm is implemented using four LSUs, three of which are used for parallel data loading and the fourth is used to store the computed result. Nine multipliers, in combination with six ALUs, are used as a two-dimensional convolution pipeline. The results from all nine

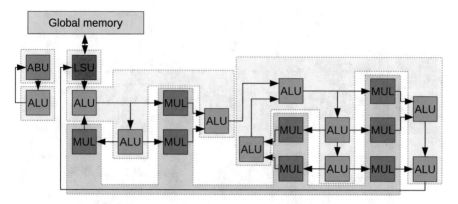

Fig. 5.11 Schematic overview of the implemented processing structure for the IIR kernel

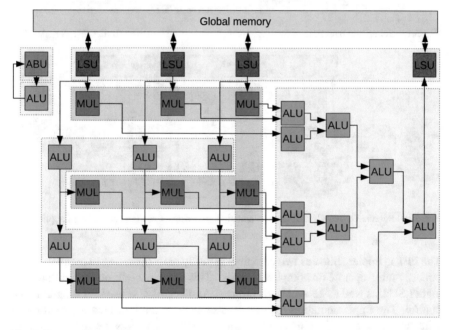

Fig. 5.12 Schematic overview of the implemented processing structure for the 2D convolution kernel

elements are summed using nine ALUs, of which one is used as a delay element. Eight instruction decoders are required to control this implementation. Figure 5.12 shows the implementation on the Blocks fabric for this kernel.

FFT   The fast Fourier transform is a method to compute a frequency spectrum from a time domain signal. The implemented FFT in this benchmark kernel is an 8-point complex FFT. Multiple FFTs are computed to produce multiple frequency spectra.

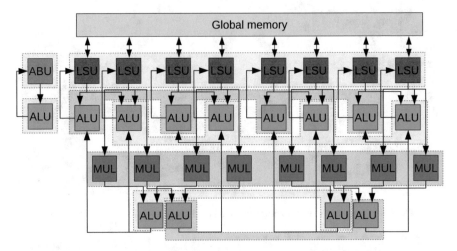

Fig. 5.13 Schematic overview of the implemented processing structure for the FFT kernel

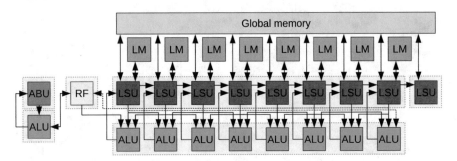

Fig. 5.14 Schematic overview of the implemented processing structure for the FFoS application

The FFT is implemented as two complete butterfly units operating in parallel. These butterfly units operate on complex numbers. This requires eight multiplier units and eight LSUs, as well as 13 ALUs to compute. Another ALU is used for program flow control. The implementation requires two immediate units to feed the multipliers with twiddle factors (coefficients). Eight instruction decoders are used to control the two butterfly units. The two butterfly units can be seen in the implementation schematic shown in Fig. 5.13.

FFoS The 'FFoS' application, a combination of binarization, erosion, and projection, is an industrial processing pipeline for Organic light-emitting diode (OLED) production. The three kernels are merged into a single application. The required resources are therefore the same as the 'projection' kernel, which has the highest resource counts of these three kernels, as shown in Fig. 5.14.

The benchmark kernels 'Binarization', 'Erosion', 'Projection', and 'FFoS' are expected to perform well on an SIMD processor since they can be relatively easily vectorized, even though the 'Erosion' benchmark does require some neighbourhood (i.e., inter-lane) communication. For the 'FIR' and 'IIR' benchmark to be efficiently executed on an SIMD, the algorithm has to be unrolled to limit neighbourhood communication. For IIR, this cannot be entirely avoided and some re-computation has to be performed. Applications that require more irregular communication (not strictly neighbour to neighbour) such as 'FFT' and 'IIR' are expected to perform better on a VLIW.

5.3 Results

This section describes the performance, power, energy, and area results for Blocks and the reference architectures. The evaluation is split into two parts. First, a comparison is made between Blocks, the reference CGRA, and the dedicated architectures to evaluate overhead. Second, Blocks is compared with the non-reconfigurable reference architectures.

5.3.1 Reconfiguration Overhead of Blocks

The goal of Blocks is to reduce reconfiguration overhead in reconfigurable archi-tectures. To evaluate the effectiveness of Blocks, application specific processors for each benchmark are compared with Blocks and the traditional CGRA in terms of power, energy, and area.

Performance

Performance, measured as the number of cycles it takes to execute a kernel, is the same for Blocks, the traditional CGRA, and the application specific architectures. This is because the underlying virtual architecture and the scheduled instructions on it are the same, see Fig. 5.15. The number of cycles can be directly compared since the architectures are all evaluated at the same clock frequency. The more densely shaded areas in the bottom of each bar represent the cycles stalled due to memory arbitration. Since the critical path is the same for Blocks and the traditional CGRA, there is no significant performance overhead from the data-path and control-path separation. In Blocks, data-level parallelism allows one instruction decoder to control multiple function units, whereas in the traditional CGRA the instructions are replicated over function units.

Stall cycles occur due to non-coalescable accesses to the global memory. In case of the binarization kernel, these are caused by eight LSUs attempting to access a byte in the global memory at the same cycle. Although the input pixels that are loaded have good locality, the memory bus is 32-bit wide. Therefore, the arbiter

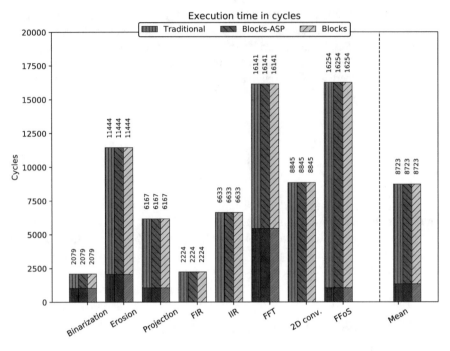

Fig. 5.15 Number of cycles for the benchmark kernels. Shaded are stalled cycles due to global memory accesses

breaks the single 64-bit wide access down into two 32-bit accesses. Each of these accesses can load four coalesced bytes. Since the binarization kernel is implemented as a single cycle loop, the overhead of the access doubles execution time. A similar access pattern occurs in the erosion, projection, and FFoS kernels. However, in these kernels, there is more data reuse through the local memories as well as more computation per data element. Therefore, the ratio of stalls to compute cycles is lower. Both the FIR and IIR kernels only use a single LSU to provide the data to these filters and, therefore, do not incur stall cycles. Similarly, although the 2D convolution benchmark uses four LSUs, only one of them performs read operations from the global memory. Two other LSUs only load from local memories, and the third only performs writes to the global memory. Therefore, full coalescing for all LSUs is not required for this kernel. The FFT kernel operates on 32-bit wide input values, which can, therefore, not be coalesced. This leads to stall cycles when loading from, or writing to, global memory. Stall cycles that occur due to non-coalescable accesses could be reduced by providing a wider memory interface. This is a power–performance trade-off and a design parameter in the Blocks framework.

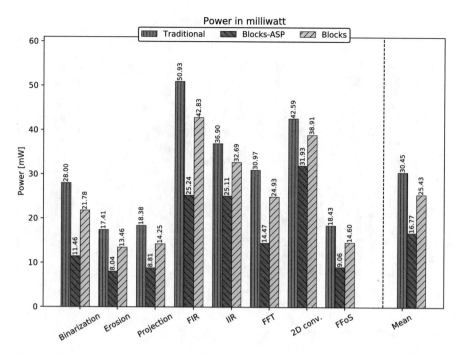

Fig. 5.16 Power per architecture per benchmark. The schedules for 'Traditional', 'Blocks-ASP', and 'Blocks' are identical and therefore have the same cycle count

Power and Energy

Reconfigurable architectures will draw more power than fixed, functionally equivalent, architectures. This is due to the extra hardware that is used to implement the reconfigurable parts of the architecture as well as function units that may not be used for certain hardware configurations. Unused function units have their inputs isolated (tied to ground) by the switch-box configuration and, therefore, mostly cause leakage. Clock gating is performed by circuitry that is automatically inserted by the ASIC tool-flow. The reconfigurable architectures for Blocks and the traditional CGRA are the same for all benchmark kernels, but their configuration differs per kernel. The effect of this can be seen in Fig. 5.16; the power per benchmark kernel varies quite significantly (from 14.25 mW for projection to 42.83 mW for the FIR filter). Power is lowest for the erosion kernel and highest for the FIR kernel. The erosion kernel uses relatively few hardware resources and has a bit lower hardware utilization than similar kernels (binarization, projection, and FFoS). The FIR kernel uses almost all resources of the reconfigurable fabric and has a very high resource utilization, and this matches with the higher power draw. It can also be observed that kernels that use multipliers (FIR, IIR, FFT, and 2D convolution) have a higher power draw than the other kernels.

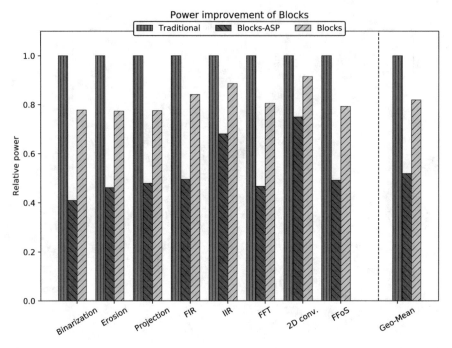

Fig. 5.17 Relative power per architecture per benchmark

The pattern is similar for the traditional CGRA and the application specific Blocks instantiations. The benchmark kernels that use more function units, and achieve a higher utilization of these units, have a higher power draw. In Fig. 5.17, it can be seen that Blocks gains less over the traditional CGRA for benchmark kernels that use more instruction decoders. This makes sense, as the effect of reusing instruction decoders within the Blocks fabric diminishes. It can also be observed that for all benchmark applications, Blocks performs better in terms of power than the traditional method. For larger designs with more parallelism opportunities, this difference will only increase further. The application specific architectures use, of course, less power than the reconfigurable CGRAs. It is important to note that these architectures use the same resources, including the same number of instruction decoders, as the reconfigurable Blocks architecture. However, they lack the reconfiguration overhead caused by the networks and unused function units. This means that the difference between the power of the dedicated architecture and the reconfigurable CGRAs is the overhead for a specific benchmark. The power overhead varies between 5.4 mW (erosion) and 17.6 mW (FIR), while the overhead for the traditional CGRA varies between 9.4 mW and 25.7 mW for the same benchmarks. Percentage-wise (compared to the corresponding application specific architecture), the overhead is between 21.9% (2D convolution) and 90.0% (Binarization) for Blocks and between 33.4% and 144.3% for the traditional CGRA.

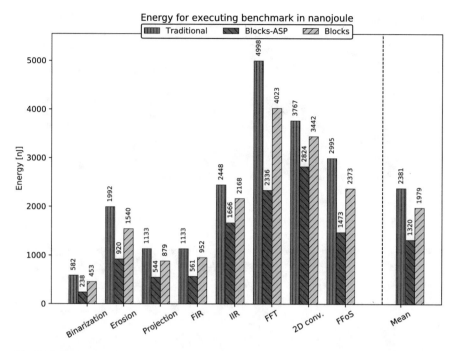

Fig. 5.18 Energy per architecture per benchmark

The average power overhead for all benchmarks of Blocks is 61% lower than for the traditional method.

Since performance for the three architectures is identical, energy differences are equal to power differences, as can be seen in Fig. 5.18, which shows the energy numbers per benchmark kernel for the architectures. Figure 5.19 shows the relative energy breakdowns for all architectures. Figure 5.19c shows that the control network of Blocks consumes between 1.1% and 8.3% (average 5.1%) of the total energy. Since the traditional CGRA does not have a control network, it does not incur this overhead, see Fig. 5.19a. However, Blocks has a much lower energy overhead in the instruction memories (on average 6.4% versus 16.5%). This is due to the data-level parallelism in the applications that Blocks can exploit by configuring vector processors on the reconfigurable fabric. This reduces the required number of instruction memories and decoders.

The overhead energy reduction between Blocks and the traditional CGRA is smallest for the 'FIR' benchmark (46% lower) and largest for the 'Projection' benchmark (76% lower). In this case, the program consists mostly of a single cycle loop that keeps the operations constant for a long period of time. This reduces toggling in the instruction memories and decoders for both Blocks and the traditional CGRA, leading to a smaller gain in energy. The 'Projection' benchmark on the other hand is a more complex kernel that cannot benefit from a single cycle loop. At the same time, Blocks can achieve a significant amount of data parallelism

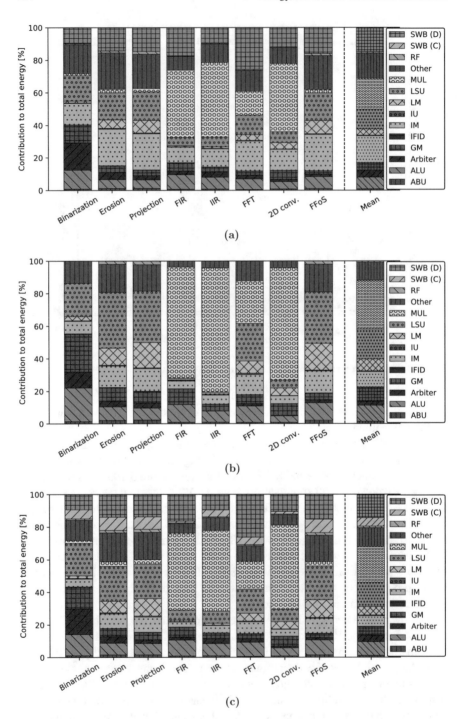

Fig. 5.19 Energy breakdown for evaluated architectures. SWB (C) and SWB (D) are switch-boxes for control and data. (**a**) Energy breakdown for the traditional CGRA. (**b**) Energy breakdown for the Blocks-ASP instances. (**c**) Energy breakdown for Blocks

in this kernel. The traditional CGRA had more activity in the instruction memories and decoders, leading to a larger overhead.

The energy fraction of the data networks is virtually identical between Blocks and the traditional CGRA. This shows that the energy reduction can be attributed to the unique control scheme of Blocks. The system level energy reduction of Blocks compared to the traditional CGRA is between 9% and 29% (average 22%). The memory macros used for all architectures are highly optimized by the manufacturer; however, our switch-boxes could still be improved, increasing the energy and area difference. First results show that a switch-box redesign, using pass transistors instead of multiplexers, reduces energy and area by 40% and 20%, respectively.

When the traditional CGRA is assumed to have a single wide instruction memory instead of multiple distributed memories, instruction memory energy can be reduced by $1.61\times$. However, as the instruction memories contribute to only 16.5% of the total energy, the reduction according to Amdahl's law is $1.07\times$. Even with this reduction, the system level energy of Blocks is still 12.1% lower. In reality, energy improvement from a single memory will be lower as instructions for all FUs will always be fetched, including unused FUs, and this might even increase energy for benchmarks that do not use all FUs.

For many reconfigurable architectures, leakage is a significant factor to the total system power. Figure 5.20 shows the leakage as a fraction of the total power for the two reconfigurable architectures as well as the application specific processors. For most benchmark kernels, the application specific processors have less leakage than the reconfigurable architectures, as expected. The exception is the 2D convolution kernel. For this kernel, the resource utilization is quite high on the reconfigurable architectures, leading to a leakage that is relatively low compared to the active power. The difference in leakage between the traditional CGRA and Blocks is very small. Depending on the benchmark kernel, either the traditional CGRA or Blocks has a higher leakage. Since the difference is so small, it is hard to determine the origin of the leakage. However, the extra network that blocks uses for control does not increase leakage significantly.

Area

The architectures for Blocks and the traditional CGRA are kept the same for all benchmarks. Only the Blocks-ASP instances have different areas as they are benchmark specific. The Blocks-ASP instances do not include reconfigurable networks and are, therefore, much smaller than Blocks and the traditional CGRA. The results in Fig. 5.21 show that Blocks does not only reduce energy overhead but also area overhead. For Blocks-ASP, the memory area (global memory, local memories, and instruction memories) for all benchmarks occupies over 70% of the total area. The remaining area is used by function units and control logic for the architecture. In Fig. 5.21, it can be observed that the application specific processors only contain the function units that are required for a kernel. For example, only the FFT, FIR, IIR, and convolution kernels need multipliers. In Blocks-ASP processors,

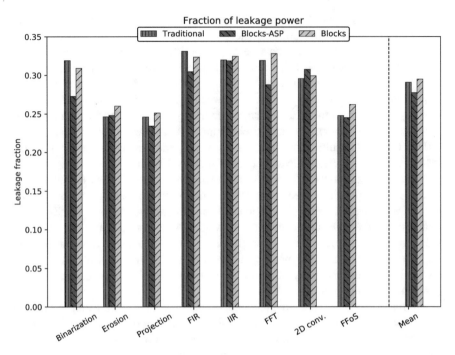

Fig. 5.20 Fraction of leakage power per architecture per benchmark

not all input or output ports are always used. Since the wiring in these processors is fixed, the unused (and therefore unconnected) ports are optimized away. This means that the used area does not always scale exactly with the number of function units. For Blocks and the traditional CGRA, these optimizations cannot be made due to the switch-box network. This means that in all overhead calculations, the optimizations that can be applied to the application specific processors are taken into account.

The area of Blocks is 19.1% smaller than the traditional CGRA. The area occupied by switch-boxes in Blocks is 49.2%, and 6.8% thereof is occupied by switch-boxes for the control network. The traditional CGRA used 46% of its area for switch-boxes, and these are only data switch-boxes as a control network is not required for this architecture. The switch-box area of the data network is slightly larger than that of Blocks. This is due to some extra connections that need to be added to distribute the program counter to all function units since these now perform local decoding. For instruction memories, Blocks uses 5.2% area, while the traditional CGRA uses 14.4% of its area. This reduction is directly due to SIMD support in Blocks.

For both CGRA architectures, a case could be made for a single wide instruction memory instead of multiple distributed memories. This would reduce the instruction memory area by approximately 1.77×. However, doing so would remove the possibility of configuring multiple independent processors on Blocks. A compromise could be found in clustering a limited number of memories for Blocks function units.

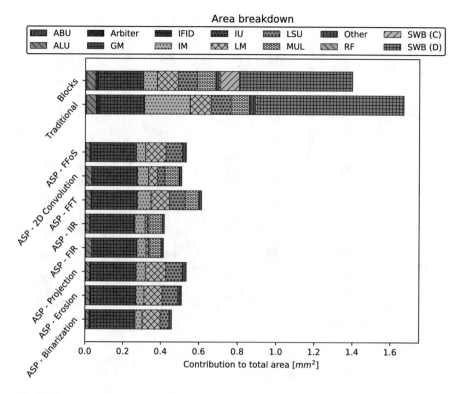

Fig. 5.21 Area breakdown for all architectures. The lower eight are Blocks-ASP instances for their respective benchmarks

Even when it is assumed that the traditional CGRA would use a single memory, the effective area reduction is only 7%. This is because the instruction memories only contribute to 14.4% of the total area. This means that the area would still be 11.6% larger than Blocks.

5.3.2   Comparison with Fixed Architectures

The real benefit of a reconfigurable architecture is that it can adapt itself to the type(s) of parallelism available in the application it is going to execute. Fixed architectures such as VLIW and SIMD processors do not have any static reconfiguration overhead. On the other hand, their supported parallelism mix is determined at design time. Blocks would show its value when it can perform similar, performance and energy-wise, compared to a VLIW and an SIMD over a range of applications. This section will demonstrate that Blocks is able to do so.

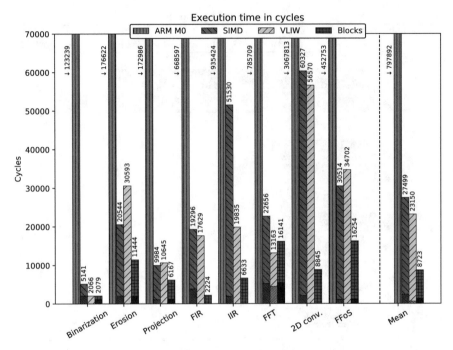

Fig. 5.22 Number of cycles for the benchmark kernels on fixed architectures. Shaded are stalled cycles due to global memory accesses. Note that the bar for the ARM Cortex-M0 is not completely shown

Performance

For each architecture, an optimized version of each benchmark has been implemented. With exception of the ARM Cortex-M0 implementations, these kernels are all handwritten and optimized. This is to ensure a fair comparison between Blocks, SIMD, and VLIW, and to not take into account compiler maturity but focus on the architecture properties. Since the ARM Cortex-M0 has a very mature compiler tool chain that can perform extensive optimizations, benchmark kernels for this architecture are written in C and compiled using GCC. The assembly output of this compiler was inspected to ensure no sub-optimal program was produced by the compiler. Figure 5.22 shows the cycle counts for each architecture and benchmark kernel. Due to the much higher cycle counts for the ARM Cortex-M0, these bars are 'cut' and the number shown next to it.

Blocks achieves similar or lower cycle counts for almost all benchmark kernels. For 'Erosion', 'Projection', and 'FFoS', Blocks is instantiated with bypass channels between the processing elements. Compared to the SIMD and VLIW, where the neighbourhood network is fixed, this saves cycles needed to move data between lanes. The benchmarks 'FIR', 'IIR', and '2D convolution' are implemented spatially. This reduces the number of instructions in the loop-nest significantly, and for

'FIR' and '2D convolution', the loop-nest becomes a single cycle. For 'Binarization' and 'FFT', Blocks requires more cycles than the VLIW. In 'Binarization', this is caused by LSU read and write address initialization (13 cycles). For 'FFT', Blocks uses its resources to construct two spatial butterfly units. The VLIW and SIMD process one butterfly per lane since further-than-neighbour communication is expensive. For 'FFT', Blocks can only use a single register file, since more RFs are not required by the other kernels. This causes a small performance penalty for FFT on Blocks but does not outweigh the energy and area penalty for the other kernels if more RFs would be added. Overall, adapting the data-path to the application Blocks provides an improvement of a factor $2.7\times$ and $3.2\times$ over the SIMD and VLIW, respectively. Kernels containing DLP perform well on the SIMD. An exception is 'Binarization', because the SIMD always loads global data with eight LSUs, causing memory stalls due to the memory bus width. The VLIW reduces the loop-nest of this kernel to a single cycle by performing a load, store, and arithmetic operation in parallel. The VLIW performs worse on kernels where the local memories are used extensively, like 'erosion' and 'FFoS', since only half of the issue slots contain LSUs. Despite the SIMD incurs stall cycles for accessing eight bytes in parallel over an 4-byte memory bus, the VLIW needs to move loaded data to neighbouring issue slots before the next memory access can be made. The SIMD performs worse on kernels with further-than-neighbour communication. 'IIR', for example, requires intermediate results from other lanes causing communication over the neighbourhood network. To prevent this, the IIR filter is unrolled, which leads to re-computation, but performs better on the SIMD. The mean performance difference between the SIMD and VLIW is small, indicating that the benchmark set provides a good balance between DLP and ILP kernels.

Overall the SIMD and VLIW, processors spend more time in data movement than Blocks does. This is due to the specialized data-paths that are configured within the Blocks fabric, whereas the SIMD and VLIW have to move the data through register files, neighbourhood networks, or local memories to the processing elements. Figure 5.23a and b shows the instruction distribution for the SIMD and VLIW, respectively. These graphs must be read with a bit of care since the lanes of the SIMD or issue slots of the VLIW may have more than one function unit inside, for example, each SIMD lane contains a register file, ALU, multiplier, and LSU. Each of these function units can perform an operation. If one of the function units inside a lane or issue slot is not performing an operation, this is registered as a 'no-operation' (NOP). This allows comparing operation distribution between SIMD, VLIW, and Blocks. It can be observed that the VLIW has to perform quite a bit more data movement for the binarization kernel than the SIMD has to. This is due to the fact that the VLIW only has four LSUs in eight issue slots. To keep all issue slots busy, the data loaded by the LSU is distributed over all these issue slots. The rest of the instruction distribution is quite similar between the SIMD and VLIW with exception of the NOPs due to stalls. As the VLIW has four LSUs compared to the SIMD with eight, it is possible, in most cases, to prevent stalls on the 32-bit global memory bus. FFT is an exception as its writes are not coalesced, due to the non-sequential access pattern of the results.

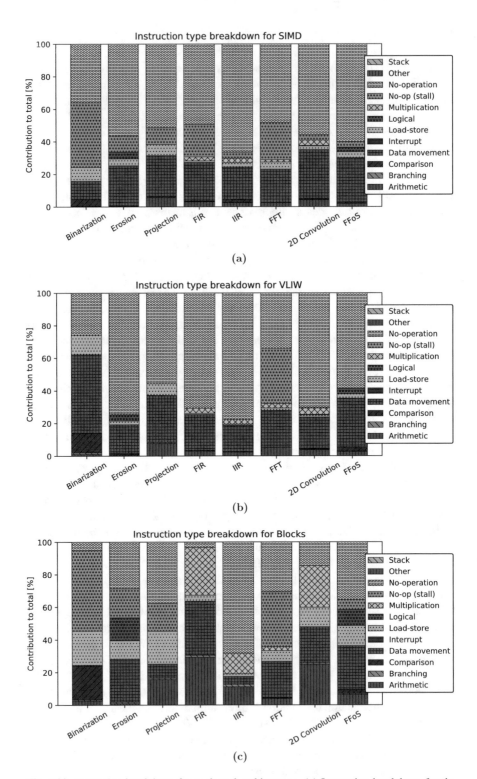

Fig. 5.23 Instruction breakdown for evaluated architectures. (**a**) Instruction breakdown for the SIMD processor. (**b**) Instruction breakdown for the VLIW processor. (**c**) Instruction breakdown for Blocks

The instruction distribution for Blocks shows much fewer NOPs in Fig. 5.23c. This is due to the fact that unused function units are isolated in configuration and, therefore, are not performing any operations (including no NOPs). Blocks uses the same number of LSUs as the SIMD processor and, therefore, the pattern of NOPs that occur due to stalls are very similar to the SIMD. Although there are still data movement operations, the operations that perform actual work such as the arithmetic and multiplication make up a much larger percentage of total operations. This is desirable as it increases the utilization of the function units. The IIR kernel has the lowest utilization of all kernels in Blocks; this is because its main loop takes two cycles per iteration to execute. During each cycle of the loop body, two-third of the function units is performing a NOP. This is due to the IIR feed-back loop that needs to be computed before the feed-forward can resume computation of the new results.

The ARM Cortex-M0 uses significantly more cycles. There are several reasons to this: it cannot perform parallel computations; control flow instructions are interleaved with computation, and there is no hardware multiplier. However, the ARM Cortex-M0 is mainly included to compare energy numbers. To predict how much difference a more powerful microprocessor makes, the cycle accurate simulator OVPsim is used to predict speed-up using an ARM Cortex-M0plus (which has a two-stage pipeline versus a three-stage pipeline in the Cortex-M0), ARM Cortex-M4f, and Cortex-A9. The highest speed-up is $2.47\times$ and $2.59\times$ for 'FFT' on the Cortex-M4f and Cortex-A9, respectively, as shown in Fig. 5.24. Speed-up is also achieved for 'FIR', 'IIR', and '2D convolution', all containing multiplications. There is almost no speed-up for the other kernels. However, core power for the Cortex-M4 increases by $2.3\times$ according to [2, 3], canceling out any energy gains. Power increase for the Cortex-A9 is even higher. The energy efficiency of a Cortex-M4f is slightly better than the Cortex-M0, but there is still a large difference with the SIMD, VLIW, and Blocks.

Power and Energy

Blocks has a higher power draw due to reconfigurability overheads. This leads to a geometric mean power increase of $1.32\times$ and $1.51\times$ compared to SIMD and VLIW, respectively. Figure 5.25 shows the power numbers for all architectures over the benchmark set. Power for Blocks is always higher because of the higher hardware utilization, and the hardware performs more work per cycle.

In most cases, the VLIW has a higher power than the SIMD. This is caused by the instruction memories, as shown in the energy breakdown in Fig. 5.26a. For kernels where the SIMD has higher power ('IIR' and '2D convolution'), the filter code was transformed to allow efficient mapping and to avoid communication between lanes, and this results in more multiplications (re-computation). Power for Blocks is highest for the FIR benchmark. In this benchmark, eight multipliers are busy for almost every cycle. A similar situation occurs for the 2D convolution kernel. In both cases a filter, either 1-dimensional or 2-dimensional, is instantiated on the reconfigurable fabric in spatial layout. As this leads to single cycle loops, hardware

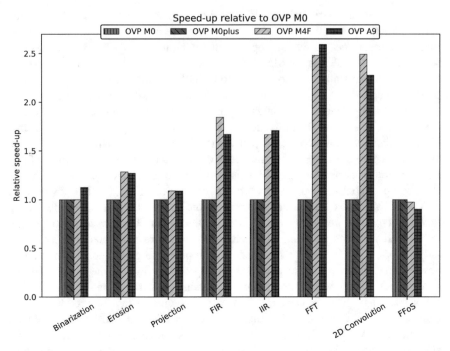

Fig. 5.24 Results of using OVPsim for various ARM processor models. The graph shows the relative speed-up in clock cycles

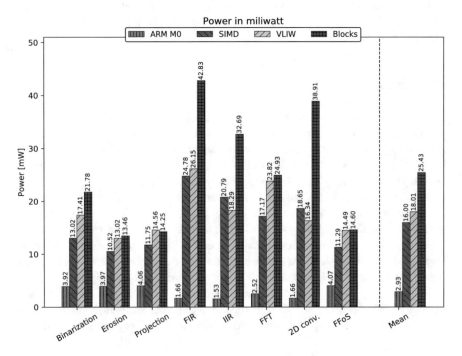

Fig. 5.25 Power results for the fixed processor architectures and Blocks

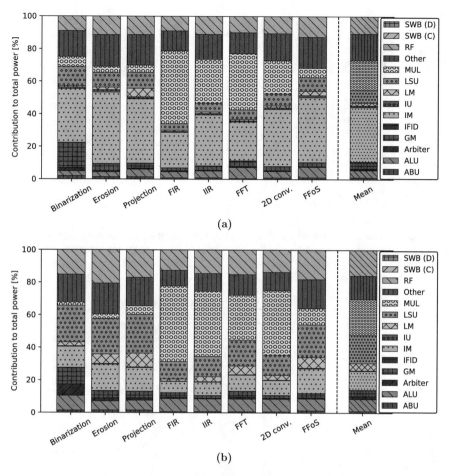

Fig. 5.26 Energy breakdown for SIMD and VLIW architectures. The VLIW processor has more instruction decoders and associated instruction memories; this can clearly be observed by the energy distribution. (**a**) Energy breakdown for VLIW. (**b**) Energy breakdown for SIMD

utilization is very high and instruction decoding and fetching overhead is very low. The ARM Cortex-M0 has a very low-power footprint, see Fig. 5.25. Interesting for this architecture is to observe that kernels containing many multiplications result in a lower power draw, this is caused by the large number of cycles required to perform multiplications with respect to the low number of memory accesses, see Fig. 5.27.

However, power is not the whole story. Since all architectures provide different performance for each benchmark kernel, the amount of energy they consume is different. Due to the performance increase, Blocks outweighs the power increase in terms of energy. Blocks shows an energy reduction of $2.05\times$ and $1.84\times$ when compared to the SIMD and VLIW, see Fig. 5.28. The cause of this can be seen when Fig. 5.26b and a is compared with Fig. 5.19c. Blocks shows a much better utilization

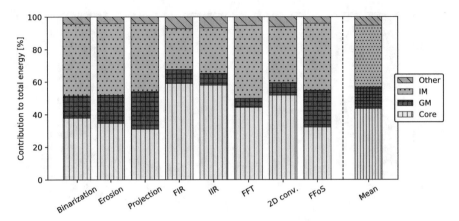

Fig. 5.27 Power breakdown for ARM M0. The energy contribution of the instruction memories (blue) and the global data memory (red) cause over half of the energy consumption. This is due to the low amount of logic in the processor core, causing a low dynamic power consumption with respect to the memories

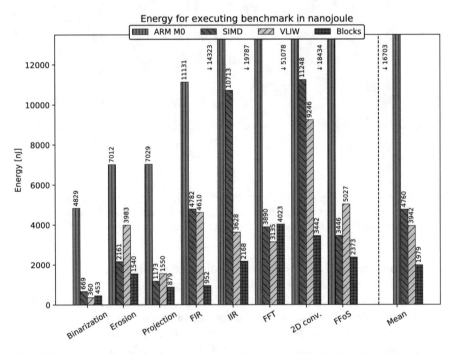

Fig. 5.28 Energy results for the fixed processor architectures and Blocks. Note that the bar for the ARM Cortex-M0 is not completely shown

of the multipliers and ALUs, leading to the energy reduction. For 'Binarization', Blocks performs slightly worse in energy than the VLIW because there is virtually no possibility for data reuse in this kernel. The VLIW suffers from the same problem but has a lower power than Blocks. For 'FFT' the number of cycles cannot be significantly reduced due to the single register file. This leads to a somewhat higher energy for the FFT benchmark, although roughly on par with the SIMD architecture. Power draw for Blocks-ASP is on par with the SIMD and VLIW architectures while outperforming these architectures in execution time by approximately 3×, leading to a significant energy reduction. This shows the gains that are possible when adapting the architecture to the algorithm.

The ARM processor, although it has a very low power draw, has a much higher energy consumption than the other architectures. This is due to the large number of cycles it takes to compute the benchmark kernels. This clearly shows that when a processor architecture does not take advantage of any of the parallelism types that may be available in applications, the energy efficiency is reduced.

Figure 5.29 shows the energy per operation for each architecture and the kernels in the benchmark set. The energy per operation is based on the peak performance of each architecture. In the case of Blocks, this is the peak performance of the virtual architecture mapped onto the physical processor fabric. The ARM Cortex-M0 always performs one operation per clock cycle. The SIMD consists of an eight-lane vector processor and a control processor, and its peak performance is therefore nine operations per clock cycle. Similarly, the VLIW can provide a peak performance

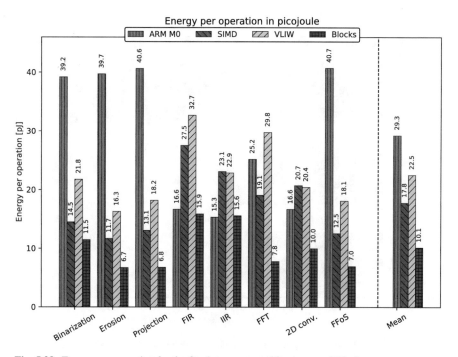

Fig. 5.29 Energy per operation for the fixed processor architectures and Blocks

of eight operations per clock cycle. The peak performance for Blocks is based on the number of used function units in the virtual architecture for each benchmark. The unused function units are not taken into account. It can be observed that Blocks provides the lowest energy per operation for all kernels with the exception of the 'IIR' kernel. The energy per operation for this kernel is slightly higher than the ARM Cortex-M0 due to lower hardware utilization.

Area

Blocks has a larger area than the fixed architectures that it is compared against. This is expected as the compute areas of all architectures are kept the same as much as possible, while Blocks has the additional area overhead of the reconfigurable networks. Figure 5.30 clearly shows the overhead caused by both the data network switch-boxes (SWB (D)) and the control network switch-boxes (SWB (C)). Since the control network has much fewer channels per switch-box, it requires much less area than the data network.

The area of Blocks without switch-box area is very similar compared to the area of the VLIW and SIMD. This shows that the compute areas of the reference architectures are well balanced, and that the extra area is causes by the Blocks

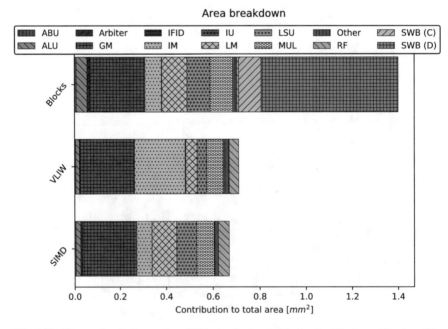

Fig. 5.30 The overhead of reconfigurability is clearly visible for the Blocks architecture. The networks for control (turquoise) and data (grey) occupy approximately half of the chip area. The area for memories and compute is comparable for all architectures

reconfigurable fabric. Important to note is that since these are post place-and-route numbers all optimizations by the synthesis tool have already been performed. This includes removal of unused logic due to unconnected ports.

The area distribution of Blocks is more similar to the SIMD than the VLIW processor, whereas the traditional CGRA, see Fig. 5.21, is more similar to a VLIW processor. This is caused by the higher number of instruction decoders. As expected, the VLIW uses more area for the instruction memories, while the SIMD has more area in the FUs. Since the SIMD has twice the number of LSUs compared to the VLIW, there is more area allocated to the local memories (since the size of each local memory is kept constant). The ARM Cortex-M0 processor is small in comparison with its memories even though the ARM Cortex-M0 memories are single ported as opposed to the two-ported memories in the other architectures, reducing their relative size.

Flexibility

The Blocks architecture aims to provide high energy efficiency and good flexibility at the same time. Section 5.3.2 already demonstrated that Blocks can provide high energy efficiency. To show that Blocks is also flexible is harder, since there is no standardized way to evaluate flexibility. Although some methods to classify flexibility have been proposed, these either base their metric on the ability of a design unit to connect to any other design unit [4] or its insensitivity to performance differences [5].

According to the metric described in [4], Blocks would always be able to connect every design unit to every other design unit, thus making it the most flexible architecture available for the four architectures evaluated in this section.

The second metric defines 'versatility' as the geometric mean of all speed-ups relative to the best performing processor for each individual benchmark kernel. Using this metric, the 'versatility' metrics can be obtained for all four architectures, as shown in Table 5.4. As the 'versatility' metric does not take performance differences into account, the ARM Cortex-M0 gets a very poor versatility rating even though it is, instinctively, very flexible. The same holds, to a lesser extent, for the SIMD and VLIW processors. The metric proposed in [5] is therefore not a good metric to quantitatively compare architectures on their flexibility when absolute performance differences are relatively large.

Since Blocks aims at energy efficiency, it seems apt to classify flexibility based on the energy spent to perform the work required to compute a benchmark kernel. The higher the variation in the amount of energy spent between benchmark kernels, the lower the flexibility. An ASIC, for example, will spend very little energy to perform a kernel that it is designed for but may take lots of energy to perform one that does not match its architecture (if it is capable to perform it at all). This means that the variation will be high. For a very flexible architecture, like a microprocessor, the variation would be much smaller as it is able to perform all applications with reasonable efficiency. However, the magnitude of the variation depends on the absolute differences between benchmark kernels. For this reason, the

Table 5.4 Versatility of the reference architectures compared to Blocks

Metric	ARM M0	SIMD	VLIW	Blocks
'Versatility'	0.02	0.31	0.40	0.97
'Energy flexibility'	0.65	1.0	0.65	0.75

Table 5.5 Summary of reference architectures compared to Blocks. Blocks provides a very similar area efficiency as the VLIW and SIMD architectures. For the ARM M0 processor, the numbers between parentheses show results only taking into account the four kernels that do not require multiplication ('binarization', 'projection', 'erosion', and 'FFoS')

Metric	VLIW	SIMD	Traditional CGRA	ARM M0
Performance	2.4×	3.1×	1.0×	68.9× (32.7×)
Power	0.76×	0.66×	1.22×	0.12× (0.17×)
Energy	1.84×	2.05×	1.22×	8.01× (7.74×)
Area	0.51×	0.48×	1.19×	0.18× (0.18×)
EDAP	2.25×	3.05×	1.45×	93.82× (45.56×)
EDP	4.42×	6.36×	1.22×	551.2× (253.1×)

energy numbers should be normalized per architecture. This method is not perfect either as it will consider architectures that have a relatively high energy overhead compared to their active energy to be more flexible. The results of this metric are shown in Table 5.4 as 'energy flexibility'. It can be observed that both the ARM Cortex-M0 and the VLIW score quite well in terms of flexibility. Both of these processors are in reality very flexible. Blocks scores somewhat less on the flexibility scale due to higher variations in the consumed energy. In reality, Blocks can be used to implement processors that are similar to both the ARM and VLIW, and this intuitively suggests a more flexible architecture.

5.4   Conclusions

Blocks performs well on cycle count and energy. Like all reconfigurable architectures, Blocks trades flexibility for area and consequently has a larger area than fixed architectures. When compared to the traditional CGRA, which has the same function units as Blocks, there is no performance improvement. However, there is improvement in both power and energy. Nonetheless, the area is smaller than the area of the traditional CGRA. In such cases, the Energy-delay-area-product (EDAP) is a common way to evaluate such trade-offs. Blocks provides a better EDAP than all non-application specific reference architectures, as shown in Table 5.5.

Another interesting metric is performance per area. For this metric, the area is divided by the number of cycles (this number of cycles is effectively the inverse of performance) it takes for a kernel to complete. Since these numbers are not meaningful by themselves, they are normalized to the performance per area of Blocks. When performance per area is considered, Blocks performs similarly or

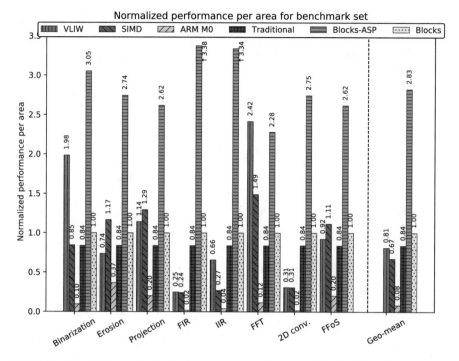

Fig. 5.31 Area efficiency results (performance per area) are normalized to the performance per area of Blocks

better than the VLIW and SIMD for most benchmarks and always better than the traditional CGRA and the ARM Cortex-M0, as shown in Fig. 5.31. The dedicated architectures have a better performance per area since their performance is identical to Blocks but all overhead is removed. There are benchmarks where the VLIW or SIMD outperform Blocks, such as the 'Binarization' and 'FFT' benchmarks. Blocks has a similar execution time for these benchmarks but requires more chip area.

For very small instruction memories and designs featuring a limited number of function units, the traditional approach may be more energy efficient. However, for typical cases and larger CGRAs, where function units often can be used in SIMD-like manner, Blocks will be more energy efficient.

References

1. Y. He et al., A configurable SIMD architecture with explicit datapath for intelligent learning, in *2016 International Conference on Embedded Computer Systems: Architectures, Modeling and Simulation (SAMOS)*, July 2016, pp. 156–163. https://doi.org/10.1109/SAMOS.2016.7818343
2. ARM, Cortex-M0 - Arm Developer, https://developer.arm.com/products/processors/cortex-m/cortex-m0. Accessed 2019 June 14
3. ARM, Cortex-M4 - Arm Developer. https://developer.arm.com/products/processors/cortex-m/cortex-m4. Accessed 2019 June 14

4. P.D. Stigall, Ö. Tasar, A measure of computer flexibility. Comput. Electric. Eng. **2**(2), 245–253 (1975). ISSN: 0045-7906. https://doi.org/10.1016/0045-7906(75)90011-7. http://www.sciencedirect.com/science/article/pii/0045790675900117
5. R. Rabbah et al., Versatility and VersaBench: a new metric and a benchmark suite for flexible architectures. December 2005

Chapter 6
Architectural Model

Designing configurations for reconfigurable architectures is not trivial and usually takes many design iterations. For each design iteration, a designer requires information on the performance, energy, and area of the configuration in order to make informed choices. The same holds for automated design tools that synthesize a high-level language description into hardware configurations and instruction schedules.

Most models and DSE frameworks for CGRAs focus at performance, some examples thereof are [1–4]. One of the most cited CGRA modeling frameworks is the CGRA-ME framework [5] which provides a tool-flow to obtain performance numbers from hardware descriptions of arbitrary CGRAs. In order to do so, an architecture is described in a custom XML based language called CGRA-ADL. From this language a software model of the CGRA is constructed that represents the physical logic and interconnect resources of the CGRA. A data-flow representation of a benchmark application is then mapped to this model to obtain performance numbers. The tool-flow allows generation of synthesizable Verilog RTL which can be used to characterize the function units. This is a requirement for every framework that wishes to integrate the Blocks-AIM energy and area estimation model.

The work presented in [6] uses efficient arrangement of reconfigurable array components and the interconnect to exploit the memory access patterns present in digital signal processing applications. Although a breakdown of area and energy is given for a baseline application as well as several benchmark applications, the proposed DSE tool does not use an area or power model to achieve these results. The power and area reductions are achieved by optimizing the memory access pattern and mapping of the application to the hardware resources. In contrast to this work there are energy aware synthesis tools [7] that do take into account energy during synthesis. In this case empirical, relative scaled, power values are used to estimate energy scaling during design exploration. Block-AIM on the other hand uses actual hardware characterization of post-synthesis net-lists to obtain absolute energy numbers. Rather than providing scaling information, Blocks-AIM provides quick but accurate absolute energy and area information for an architecture instance.

M. Wijtvliet et al., *Blocks, Towards Energy-efficient, Coarse-grained Reconfigurable Architectures*, https://doi.org/10.1007/978-3-030-79774-4_6

Recent work [8] uses micro-benchmarking to train a machine learning model. This model is subsequently used to analyse application performance. A similar approach could be used for energy and area as well. However, if inaccuracies in the final estimations do occur it is much harder to determine the cause of these deviations. Therefore, Blocks-AIM uses an analytical method to estimate energy and area.

Power and path delay estimation has been done for FPGAs [9]. In this work, system power is estimated based on the utilization levels of interconnect and multiplexer resources. The results of this work show a mean error of 21%. However, the overhead of the interconnect is much lower for CGRAs compared to FPGAs. Due to this, the interconnect and multiplexer utilization may not be a good basis for CGRA energy estimation.

ASIC design tools also perform power estimation. Traditionally, this is done based on a library of characterized gates. This information is then directly referenced to estimate the power for an IP Block. Another way of estimating power for IP blocks is to use a power contributor model [10]. In such a model the physical parameters such as switching capacitance, resistances, and current sources are stored instead of extracted power numbers. This has the advantage that the model is frequency, process, voltage, and temperature. Another method is to use a hybrid power model [11] that uses transistor-level information at gate level. This helps to increase the power estimation accuracy but does not require electrical simulations. Both of these methods significantly increase power estimation with respect to traditional simulations. These models achieve this by abstracting from the fine-grained building blocks. The proposed method uses a similar technique but at a higher abstraction level. Instead of abstracting from transistors to gates the model abstracts from gates to CGRA building blocks, such as function units.

The method presented in this paper does not only hold for the Blocks reconfigurable architecture. The function unit characterization method can be used for many different CGRAs such as KAHRISMA [12], TRIPS [13], ADRES [14], and RAW [15], as long as RTL implementations of these architectures are available or can be generated.

Performance can be obtained by simulation of the designed instruction schedule. As Blocks is predictable this will yield an exact cycle count. The predictability of Blocks also allows for execution time analysis by automated design tools as long as the execution time is not data dependent.

Energy and area estimation usually requires at least a complete logic synthesis run. For a Blocks design, as used in Chap. 5, this takes over two hours to complete. Although these energy and area estimations are very accurate, taking two hours for each design iteration is far too long when many iterations have to be performed. It is possible to use pre-synthesized macros for building blocks such as function units and switch-boxes. This will improve the synthesis speed significantly. However, this may come at the expense of a less optimized system level design as no cross module optimizations are made. Furthermore, it is still required to perform RTL-level simulation and power extraction to obtain energy consumption. This still takes a significant amount of time. Using a model for energy and area estimation will be less accurate but several orders of magnitude faster. Even though a model may be less accurate in terms of absolute energy and area numbers the way a model scales for small design changes is usually accurate enough for Design space exploration (DSE).

This chapter introduces an energy and area model for Blocks that allows fast exploration of architecture properties. Section 6.1 introduces how energy and area information is obtained for the Blocks architecture. Section 6.2 then shows how this information can be used to construct an energy and area model. Finally, Sect. 6.3 provides an evaluation on the accuracy of the model. Finally, Sect. 6.4 concludes this chapter.

6.1 Micro-benchmarks

Energy and area models require some basic information about the properties of an architecture. These properties can be determined by performing micro-benchmarks on individual building blocks of an architecture. Since Blocks has a relatively limited number of building blocks (function units, switch-boxes, instruction decoders) it is feasible to individually profile these units as a one-time effort and obtain energy numbers for the operations these building blocks can perform. Based on the properties of the individual building blocks the total energy and area of the architecture can be estimated.

Obtaining accurate energy numbers for the individual building blocks requires applying known inputs to the Design under test (DUT). This requires a test-bench which, after synthesis of the DUT, applies a sequence of inputs while an activity trace of all internal nets of the DUT is recorded. Within the Blocks framework the test-bench for most DUTs could be formed by a slightly modified version of a hard-wired Blocks instance. This instance contains some function units for execution control (an ABU, ALU, and immediate unit) as well as a test-bench controller. The function units load an application that controls their operation, and with this the program flow. The test-bench controller is a sequencer that provides inputs to the DUT synchronously with program execution. Figure 6.1 shows the structure of the design used for unit testing, only the 'compute module' is synthesized. Some DUTs, such as the switch-boxes, have their own dedicated test-benches as this allows for better direct control.

This section describes how the required input vectors and application are generated, how the energy and area numbers for function units are obtained, and how energy and area for the memories are estimated.

This process is almost fully automated. The only manual step that is required is to specify a configuration for the test-benches used for characterization. In this configuration, the instructions to be characterized, their number of operands, and supported data types are specified. This has to be done once for each function unit type. In the future, this step can be automated, as the data is available within the Blocks framework. However, as this needs to be done only once for a few function

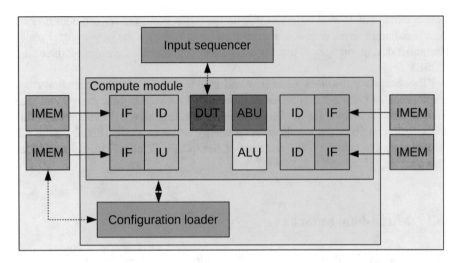

Fig. 6.1 Unit test structure for function units. Shown are the compute module and the memory module containing the instruction memories. The device under test (DUT) is shown in red and represents the function unit that is being evaluated

units this was performed manually. The remaining steps are fully automated. For each function unit the configuration file is processed, the hardware description is synthesized, and the results are gathered and parsed for storage in the model database.

6.1.1   Input Vector and Program Sequence Generation

In digital circuits, the energy consumption depends strongly on the switching activity in the circuit. We take an inverter as an example. Every time the input value changes (toggles) the output of the inverter will also change. When this happens the capacitance of the input gates of the fan-out gates has to be either charged or discharged, thus causing a current to flow through either the NMOS or PMOS transistor. The input value of the inverter does not change instantly but has a slope. This means that there is a point at which both transistors are partially conductive, thus causing a current to flow from the positive to the negative power rails of the circuit. Due to these properties CMOS circuits have their highest power draw when switching. This is called dynamic power dissipation and is data dependent. In the steady state the circuits still have some leakage through the transistors. This is called static power dissipation.

The expectation is that energy numbers for the DUTs vary significantly with the number of toggling input bits. However, the model is expected to operate on an instruction trace only. Therefore, no information about the actual data passing through the design can be assumed. This would be the case in a Design space exploration (DSE) tool that is iterating over an instruction schedule and a hardware configuration, in this case the operations are known, but there is no information about the data values. The only assumption that can be made is that, most likely, the data values loaded by the LSUs are a reasonable representation of the data-width in the data-path.

For this reason, the DUTs are evaluated for the data-widths supported by the Blocks LSUs: double-word (64-bit), word (32-bit), half-word (16-bit), and byte (8-bit). Of course, this is limited by the configured data-path width of the design. Throughout this chapter profiling will be performed using 32-bit function units, because the results thereof can be directly compared with the results obtained for Blocks in previous chapters. This requires DUT evaluation for word, half-word, and byte values. For a byte this means that the lower eight bits will be applied to the DUT with a randomly generated value while the upper 24 bits will be kept constant at zero.

The input values are generated as (uniform) randomly toggling individual bits for a specified width for each of the inputs of the DUT. These input values are stored as a sequence of input vectors, the input sequencer in Fig. 6.1 provides these inputs synchronously with a Blocks application. If the DUT is a function unit, the application is generated in parallel with the input vectors. This application contains all operations valid for the DUT. Most function units have multiple inputs and outputs, the test-bench activates all possible input and output combinations of these units for at least a few hundred times. This is done in order to obtain a reliable representation of the energy used by each operation. Some units, such as the ABU, can operate in multiple configurations (e.g. branching mode and accumulation mode). Since this would require reconfiguration half-way the test sequence these configurations are evaluated as independent tests.

The switch-boxes require a different structure. Various switch-box configurations are loaded, one after the other. Each configuration implements a number of configured paths through the switch-box, this allows to obtain energy numbers for different numbers of configured paths. This allows for interpolation and extrapolation in the energy model. Each configuration is exercised with multiple input vectors, similar to the function units.

6.1.2 Obtaining Energy and Area Numbers

Typically, synthesis tools report only average power over a whole simulation run, which would require one simulation run per operation that needs to be classified. The Joules application within the Cadence ASIC design tool-chain can provide power per time interval. Setting this time interval to one clock period provides per cycle power measurements. Each clock-cycle, therefore, becomes a measurement 'segment' for Joules. Unfortunately, Joules is limited to 1000 segments per stored activity file. It was possible to work around this limit to store multiple activity files per simulation run, one for every 1000 clock cycles.

Obtaining power per cycle instead of per operation has the advantage that power can be correlated to the number of changing input bits. Although not used in the current energy model it can be used to gain more insight into why a model may be off in some situations. For each evaluated DUT, and for every operation it can perform, at least 1000 measurements were performed for each data type. By doing so, an accurate distribution of the power draw can be obtained. Furthermore, the actual number of toggling bits is correlated such that it is possible to estimate power draw based on the number of actual toggling bits.

After power numbers are obtained, these numbers need to be related to the evaluated operations. This is performed by a script that takes a trace of these operations and matches them to the corresponding cycle in the power estimation. Since the clock period is known, the power numbers can be converted into energy per operation. These numbers can then be used to gain insight in the behaviour of the DUT via plots such as those shown in Fig. 6.2. This figure shows the energy per operation for the ALU. It can be observed that within the same data size the medians are quite similar, with the exception of operations such as 'add' and 'sub'. These operations cause rippling in the full-adders inside the ALU which cause a somewhat higher energy number. The 'nop' operation is very noticeable due to its much lower median combined with its very skewed distribution. Although the median is around 0.2 pJ per operation there are numbers up to 2 pJ per operation reported on the higher side of the distribution. These are caused by toggling the input values even

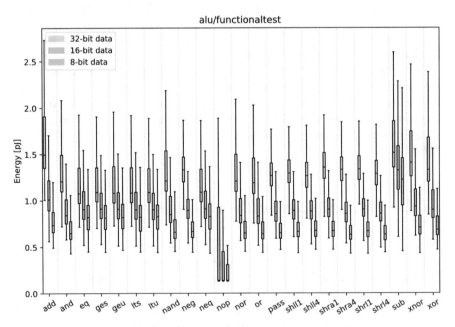

Fig. 6.2 Energy per operation at various data-widths as recorded for the ALU function unit, for a 40nm low-power technology library at nominal voltage and typical temperature

though the ALU does not write to any output register. The changing input values cause changes in the combinatorial logic which cause some power draw. Actively isolating unused inputs may prevent this but causes higher energy for all operations, as well as a longer critical path.

This profiling method works well for most function units. However, the multiplier is more sensitive to the actual data input, as can be observed in Fig. 6.3. In this case, the median values are much further apart than those shown for the ALU, in Fig. 6.2. For the multiplier it is therefore important that at least the data type is used in the model. Without it, the error will become significant. Especially since the multipliers are likely one of the larger, and more energy consuming, building blocks in CGRAs. It is also interesting to observe that also for pass and no-operations the energy consumption is still significant. The reason for this is similar as for the ALU, there is still toggling activity in the combinational part of the circuit. The multiplier is characterized for word, half-word, and byte data types. Due to the sensitivity of the multiplier to the actual data it may be advantageous to profile intermediate bit-widths, such as a half-byte. However, this requires knowledge of the actual data

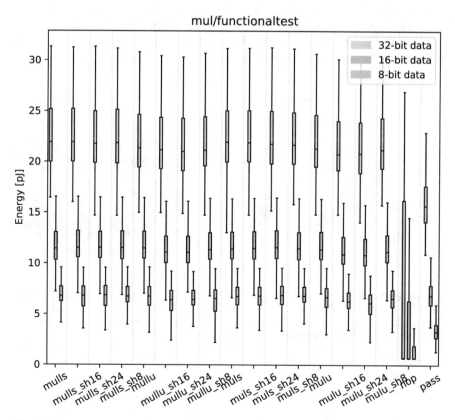

Fig. 6.3 Energy per operation for instructions at various data-widths as recorded for the MUL function unit

values during execution, something we do not know. If the compiler would be able to estimate the data range or toggling of the data values this would help to reduced the energy estimation error. For this model, however, we assume that these parameters are not known.

The extracted energy information is used to construct a database of information about each DUT. The database includes, for each operation, the median, mean, fitted normal distribution, and fitted skew-normal distribution for each of the operations. The fitted probability distributions were added to the database as it was uncertain if an energy model based on just the median, or the mean, would be sufficiently accurate. The probability distributions for two ALU operations are shown in Fig. 6.4. The green histogram represents measured values, the blue line a fitted normal distribution, and the black line a fitted skew-normal distribution. It is clear that the skew-normal distribution is a better choice to represent the energy behaviour of the operations. It can be observed that the 'add' operation has a secondary 'bump' near at the higher end of the distribution. This bump can be observed for multiple operations and occurs for two's-complement values when the sign changes, causing several bits to toggle.

Extracting area for each DUT is significantly easier as it does not depend on changing input values and is the same for all operations. The area is reported after synthesis and is extracted by the database generation scripts. These numbers can then be used for area estimation with the model.

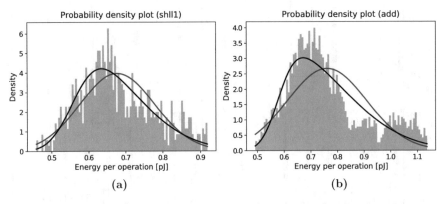

Fig. 6.4 Probability distributions for two ALU operations with homogeneous random data input. The green histogram represents measured values, the blue line a fitted normal distribution, and the black line a fitted skew-normal distribution. (**a**) Probability distribution plot for the 'shift left' operation. (**b**) Probability distribution plot for the 'add' operation

6.1.3 Energy and Area Estimation for Memories

Memories used for Blocks are based on compiled memory macros designed for a specific ASIC technology. In this case a 40nm commercial technology library is used. For such technology libraries datasheets are provided which specify power per read or write access, leakage, and area. An alternative is to use the CACTI model [16], but this model is intended for larger memory sizes, and it is inaccurate for very small memories, like those used in Blocks.

Instead, a script extracts the relevant values for all reported memory widths and depths from the provided memory datasheets and stores these in a database that can be used by the energy and area models.

6.2 Energy and Area Models

The goal of the energy and area models is to provide a reasonably accurate estimation of the energy and area for an architecture and instruction trace in a short amount of time. This last requirement is important for automated DSE as the model is in the critical path of the design iterations.

The energy and area models presented in this chapter are calibrated on the Blocks architecture, but the same methodology can, in principle, be adapted to various other coarse-grained reconfigurable architectures (CGRA). The proposed model takes an application trace, or instruction schedule, as well as an architecture description as an input. Based on this information an energy and area estimation will be made. The model is intended to provide information for both hardware resource design space exploration as well as energy aware compilers. In such cases, no cycle accurate simulation will be performed and the actual data that is processed is therefore unknown. This requires the model to be data agnostic. This will introduce some error but allows energy estimation for (partial) instruction schedules or hardware changes without fully simulating these in advance. In order to keep the estimation errors limited, the data types used in the schedule are used to improve model accuracy.

The database with extracted energy profiles, described in the previous section, allows calculation with either the median values or with skew-normal distributions. However, although computing with probability distributions can give more insight in the range of energy consumption that can be expected, the median of the resulting distribution will still be the same as computing energy consumption directly with medians. As a first approximation, and to limit computational complexity, the model described in this section operates based on the median energy results for each building block and its operations. However, it could be extended to operate on distributions if required.

Evaluation of the model is automated, as a Python script. It reads the model database, processes the trace, and computes the results. The model is integrated into the Blocks framework. Once the user has designed an application, the model can be called with a single command. The goal is to integrate the model with the design space exploration tools that are currently in development.

6.2.1 Model Requirements for Design Space Exploration

For CGRAs there are two aspects of an architecture instance that can be modified at design time. The first is the hardware structure, e.g. which function unit types are present in the design and how they are connected. The aspect that can be modified is an application that is scheduled on the hardware by a compiler. These two processes are not independent from each other. For example, during instruction scheduling a bottleneck becomes apparent on a specific function unit. The compiler could then decide to use another function unit of the same type. However, when the goal is energy efficiency this may not be the best choice. Such a decision can only be evaluated when an estimation can be made of the cost of adding another unit. This can be done with a model that estimates the energy change. Although it is desirable to have a model that provides accurate absolute numbers, it is more important that the scaling is correct. For example, when doubling the number of function units we would expect the power consumption to roughly double.

The cost functions for DSE tools usually operate on scalar values, not on probability distributions. For that reason the median for a certain data type in the model database is used. The median value was chosen over the average as it is less biased towards outliers. As can be seen in the box plots in Figs. 6.2 and 6.3, the outliers are not always evenly distributed on both sides of the box. In case the worst case energy consumption is desired it is possible to take the maximum instead of the median. However, for DSE where the model is used to make decisions based on scaling, this may lead to unwanted effects when the variation is very large.

As the goal is to evaluate (partial) instruction schedules, it is not always possible, or desirable, to perform a cycle accurate simulation that includes actual input values. In other words, the actual data values are unknown. However, the data type is known. For example, we know if a word or byte is used to represent a variable. These properties lead to the following requirements on the model:

1. A model that provides reasonable absolute numbers and accurate scaling properties.
2. The output of the model must be usable as an input for DSE tools. For example, a scalar value that can be used in a cost function.
3. The model shall be data agnostic, as actual data values may not be known during DSE.
4. The model must provide a breakdown of area and energy consumption by function unit type, allowing targeted optimization during DSE.

6.2.2 Energy Model

The energy model takes the architecture description, the application code (PASM), and a trace of the program counter as an input. Another input for the model is the database with energy related properties for each building block, as shown in Fig. 6.5. This database is ASIC technology dependent. This means that for each of these technologies, or specific temperature and voltage corners thereof, the Blocks energy, area, and timing properties need to be evaluated as a one-time effort.

The architecture and application code are part of a Blocks configuration. A program counter trace can be obtained either by simulation or based on polyhedral models such as Polly [17]. The trace is required by the energy model to evaluate how many times each branch is taken. In other words, how many times each loop is executed. The energy model follows the program counter trace and looks up which line of instructions is executed in the application code. If there is an LSU operation that performs a load or store on one of the memories, then the first argument, which specifies the data type that is loaded, is evaluated. The assumed data-width for all instructions afterwards, up to the next load or store, are then assumed to operate on this width. Although this is not as accurate as performing a simulation and generating a trace of the actual encountered data-widths it allows the model to be used without any assumption on the data other than the information provided in the instructions. This is typically also the case in compilers; the actual data values are often not known but the data types of their variables are. The total energy for a Blocks architecture instance is computed by Eq. 6.1.

$$E = E_{array} + E_{arbiter} + \sum_{\forall memories} E_{memory} \qquad (6.1)$$

Where E_{array} indicates the energy consumption in the (reconfigurable) array, this includes the function units, instruction decoders, and switch-boxes. The energy consumption of the memory arbiter is described by $E_{arbiter}$, and the energy for each of the memories (local and global) is indicated with E_{memory}. As will be shown in the

Fig. 6.5 Inputs and outputs of the energy model

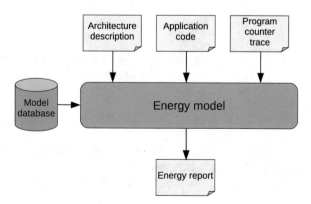

remainder of this section, each of these terms can again be broken down in smaller parts.

Computing Energy for Function Units

While traversing the instruction trace the occurrence of each operation, and the detected data-width, is counted. This is done for each individual instruction decoder. Each decoder has one or more function units under its control. In case of the hard-wired CGRAs the number and type of the function units can simply be looked up in the architecture description. However, in case of a reconfigurable architecture there is a grid of function units that forms a certain architecture instance through their configuration. To determine which function units are under control of an instruction decoder, the paths routed over the switch-boxes have to be analysed. This is performed as a preprocessing step in the energy model.

In reconfigurable architectures it is very well possible that not all function units are used. In Blocks, unused function units have their data inputs and control inputs tied to a zero value. This results in these units effectively performing a no-operation. Additionally, the ASIC tool-flow performs automatic clock gate insertion.

For each instruction decoder and its associated function units, the energy for every operation is added to form the total energy consumed, see Eq. 6.2. This includes the unused units which are assumed to perform no-operations. In this equation, E_{ID} indicates the energy consumed by the instruction decoder for each performed operation. This instruction decoder can control multiple function units as an SIMD, this number is indicated with N_{FU}. The energy consumed in a function unit depends on the executed operation and on the datatype of the operands. This is indicated with the function $f_{FUenergy}(dtype, oper)$ which essentially performs a look-up in the model database. To compute the final result for the (reconfigurable) array, the energy dissipated by the switch-boxes still needs to be added, which is indicated with $E_{switchboxes}$ in Eq. 6.3. For hardwired architecture instances, the switch-box energy does not have to be added.

$$E_{compute} = \sum_{\forall operations} E_{ID} + f_{FUenergy}(operation, datatype) \cdot N_{FU} \qquad (6.2)$$

$$E_{array} = \sum_{\forall compute} E_{compute} + E_{switchboxes} \qquad (6.3)$$

Computing Energy for Switch-Boxes

The energy consumed by the switch-boxes depends mostly on the number of paths through the switch-box that has toggling data lines. For the switch-boxes in the control network this depends on the paths from instruction decoder outputs to function unit instruction inputs. These paths are determined by recursively following the output of each instruction decoder through the switch-boxes to their destinations. By doing so a total number of 'activated' switch-boxes can be determined for each instruction decoder output change. Every operation other than a no-operation (nop) is considered a change on the instruction network.

Figure 6.6 shows an example network of three function units and six switch-boxes. The arrows indicate configurable paths between switch-boxes and function units. The red and blue paths are configured paths that can be activated if an operation other than a 'nop' is executed. For each switch-box the number of configured paths is indicated with $P_{x,y}$, this is the maximum number of active paths through that switch-box at horizontal location x and vertical location y. It can be observed that for the switch-box with $P_{1,1}$ the number of paths is three. This is because the red path splits to provide data to two separate outputs. Effectively, each used output is considered to be a path as it requires configuration (and indeed a path) through the multiplexers that make up the switch-box.

The producer of data for the red path is FU_1, the producer for the blue path is FU_3. In case that FU_3 executes a no-operation and FU_1 an 'active' operation, only the red path is considered activated. In this case the number of active paths is two for the switch-box at location $(1,1)$. Based on the energy numbers for the benchmarked switch-box, the energy is predicted for the switch-box in the architecture instance. In case the exact number of active paths is not in the model, the number is interpolated or extrapolated depending on the available data in the model.

In reality, the energy consumed by a path through the switch-box depends on the actual number of toggling bits. However, to keep the model relatively simple, and not require a data trace of all data on the network, the data type is assumed to have the width inferred from the load and store operations, similar to the energy calculation for the function units.

The switch-box energy can be calculated with Eq. 6.4. The function f_{energy} performs a look-up in the model for a given number of active paths (P_{active}) and type of data passed over this channel in a given cycle ($datatype$). If required, this function performs interpolation or extrapolation. Since switch-boxes have some static power dissipation, meaning that there is energy consumed even if the switch-box is idle. This is captured with E_{idle}.

Fig. 6.6 Example showing active paths through the switch-boxes. The red and blue lines are paths that can be activated with operations other than a no-operation

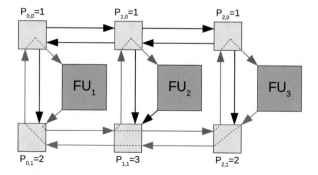

$$E_{switchboxes} = \sum_{\forall switch-boxes} f_{energy}(P_{active}, datatype) + E_{idle} \qquad (6.4)$$

Computing Energy for the Arbiter

The arbiter that handles access control for read and write requests issued by the LSUs is instantiated with a number of request ports equal to the total number of LSUs in the design. Various arbiter sizes are evaluated during benchmarking. If an arbiter size is not present in the model database, the energy numbers for a certain operation are interpolated between the two closest arbiter sizes, or extrapolated if the arbiter size is larger than those available in the model database.

The number of cycles it takes to complete all access requests depends not only on the number of requests but also on whether the requests can be coalesced into fewer memory operations. As the model performs no data analysis it cannot tell, only based on the operation issued by an LSU, whether an access is going to be coalesced or not. However, when performing simulation the program counter trace shows how many stall cycles occurred after each set of access requests. A compiler, which has knowledge of the access patterns, could provide the same information. The energy consumption for the arbiter (which is active even during stall cycles) is slightly different for stalled and non-stalled cycles. This is due to the initial detection which access requests can be performed in parallel. Figure 6.7 shows the energy per operation for an arbiter with four request ports. Some operations are shown multiple times: for uncoalesced (u), coalesced (c), or broadcast (b) accesses. Operations that caused the arbiter to insert a stall cycle in the processing pipeline are marked with (s). There is some difference in energy between coalesced en uncoalesced memory accesses. But since it is now known in the model whether accesses are coalesced, the assumption is made that the programmer or compiler will attempt to achieve coalesced accesses where possible.

In Fig. 6.7 it can be observed that, for the arbiter, the data type does not make much difference to the energy numbers. This is the case for all evaluated arbiter sizes and is likely caused by the fact that the data loaded from the memory contains non-zero bits over the full data-width. The active bits for each data type are only selected at the LSU, and not yet in the arbiter. Since the energy numbers are so similar, the energy model of the arbiter does not take into account the data type of these operations. This leads to Eq. 6.5 for estimating arbiter energy.

$$E_{arbiter} = \sum_{\forall operations} f_{energy}(operation) + E_{idle} \qquad (6.5)$$

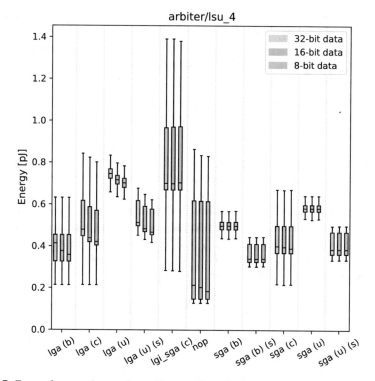

Fig. 6.7 Energy for operations performed by an arbiter with four request ports. It can be observed that the data type is relatively unimportant

Computing Energy for the Memories

The memory energy numbers are directly based on the manufacturer datasheets for memory macros available for the used ASIC technology. In the Blocks architecture two memory types are used: dual ported register files (for local memories and instruction memories) and dual ported SRAM (for the global memory). In both cases the memories are the so-called simple dual ported memories, where one port can only read and the other port can only write. Both memories have enabled inputs for both the read and write ports.

To estimate memory energy, the accesses to each of the memories are counted. For the global memory this consists of counting global load and global store operations. These take at least one cycle but may take more in case multiple memory accesses are required. Therefore, the number of memory accesses is counted as shown in Eq. 6.6.

$$N_{GMaccesses} = \sum_{\forall GMoperations} (1 + N_{stalls}^{GM_{operation}}) \qquad (6.6)$$

Where N_{stalls} is the number of stalls caused by the global memory operation (*GMoperations*) issued to the arbiter. Reads and writes are counted separately to compute energy, but using the same equation. Although the processor is inactive during a stall cycle, there are still memory operations pending. For this reason stalls are counted as memory accesses.

For local memory accesses there is no arbiter between the LSU and the memory. Therefore, no stalls occur and the number of accesses is equal to the number of times a local load of local store operation is performed. For the instruction memories, which currently operate without a loop buffer, the number of read accesses is equal to the length of the instruction trace. Although there is no instruction decoded when a stall occurs, the instruction memory ports are currently not disabled in this time period.

With the number of accesses known, the energy for each memory can be computed using the energy numbers in the energy model database according to Eq. 6.7. In case the exact memory type is not present in the database, either in the number of entries of in the width of the memory, interpolation or extrapolation is used to estimate the energy numbers.

$$E_{memory} = (N_L \cdot E_L) + (N_S \cdot E_S) + (N_{notIdle} \cdot E_{static}) + (N_{idle} \cdot E_{idle}) \qquad (6.7)$$

The energy for loads (E_L) and stores (E_S) are different and are multiplied with the computed number of loads (N_L) and stores (N_S). On top of the access energy there is also some static energy overhead. Furthermore, when the read and write ports of the memory are disabled (when the memory is idle) there is still a leakage current that causes energy consumption when no loads or stores are performed (E_{static}). This current is lower than the static power dissipation for enabled memories and is, therefore, counted separately. The idle energy for an operation is multiplied with the number of cycles that the memory is idle (N_{idle}). Unused memories, for example, local memories of an unused LSU, are considered idle for the length of the instruction trace.

6.2.3 Area Model

The area model is much simpler than the energy model since it is not required to obtain any usage information on the function units or the operations they perform. The area model, therefore, only requires the architecture description to obtain information about how many function units, memories, and switch-boxes are present. Figure 6.8 shows the flow to obtain area numbers.

The array area can be obtained by simply adding up the area for the number of function units (N_{type}) of each type (*type*). The function $f_{area}(type)$ performs a look-up in the model database for the area of each function unit type. If present, the area (A_{swb}) for the switch-boxes in the control and data network is added to the array area, multiplied by the number of switch-boxes in each of these networks

Fig. 6.8 Inputs and outputs of the area model

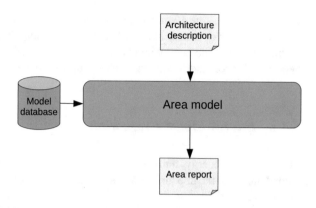

(N_{swb}). This leads to Eq. 6.8.

$$A_{array} = \sum_{\forall type} \left(f_{area}(type) \cdot N_{type} \right) + A_{swb}^{data} \cdot N_{swb}^{data} + A_{swb}^{control} \cdot N_{swb}^{control} \tag{6.8}$$

The area for the arbiter depends on the number of request channels, this number is equal to the number of LSUs present in the design. If no arbiter is present in the model database that has the same number of request channels, then the area is interpolated or extrapolated to estimate the actual arbiter area.

Memory area is, similar as the memory energy, obtained from the memory datasheets. These numbers are (of course) provided for a finite set of memory configurations (depth and width). This often requires interpolation or extrapolation on the depth, width, or both to obtain an estimate for the memory area. For example, the datasheet for the dual ported SRAM does not provide information about memories with a 32-bit word width. Instead it provides data for memories with an 18 or 36-bit width.

The total area for a Blocks architecture instance is computed according to Eq. 6.9. This ignores the area for smaller modules in the design, such as the configuration loader and DTL interfaces. However, the total area of these is comparatively small to the area of even a single function unit.

$$A = A_{compute} + A_{arbiter} + \sum_{\forall memories} A_{memory} \tag{6.9}$$

6.3   Evaluation

For the model to be usable for DSE it needs to predict energy and area numbers close to the actual numbers. In this section, evaluation of the model is performed. As a reference, the numbers reported by the ASIC synthesis tool are used. These numbers are based on the same 40nm technology library as used for micro-benchmarking the function units. For evaluation, the benchmark kernels from Chap. 5 are used. This section evaluates the energy and area models in Sects. 6.3.1 and 6.3.1, respectively. Section 6.3.3 concludes with an overview of the execution times of the model.

6.3.1   Energy Model Evaluation

The reference energy numbers are based on the post-synthesis net-list of the architectures for the benchmark kernels. RTL simulation is performed on the net-list to obtain an activity file. This file contains the toggle rates per net over a specified time period. The time period starts when the Blocks architecture is configured and starts processing, and it ends when the benchmark kernel completes execution and halts the processor. The activity file, together with the net-list, is used for power estimation using tools in the ASIC tool-flow.

Energy Estimation for Hard-Wired Instances

Figure 6.9 shows the reference and estimated energy numbers for Blocks-ASP instances. These instances are hard-wired CGRAs (without switch-boxes) that essentially form application specific processors. Each benchmark kernel has its own hard-wired architecture, generated using the Blocks framework. It can be observed that for all benchmark kernels except '2D convolution' the predicted energy numbers are very close to the reference numbers. Not taking '2D convolution' into account, the error of the model lies between -12% for 'Binarization' and $+4\%$ for 'FIR'.

The energy for the '2D convolution' benchmark is 39% overestimated by the model. This is due to the data type used in the application in relation to the actual values. The benchmark performs a 3x3 convolution on an image. The values of this image are loaded as bytes and range between 0 and 255, using the entire range of the 8-bit data type. However, the convolution coefficients range between 0 and 2. Furthermore, the convolution is implemented as a spatial layout using nine multipliers. The convolution coefficients are, therefore, entirely static during program execution. The energy estimation in the model is based on a multiplication of 8-bit by 8-bit values, while in reality only 8-bits can toggle in this benchmark application. Figure 6.3 shows the cost of different multiplications for a varying number of bits. For the specific multiplication performed in this benchmark (mullu_sh8) the median energy is 6.6 pJ per operation. But since most of the coefficients are either zero or one the actual energy value is closer to a 'pass' operation. Such an operation simply forwards the input value to an output. The energy median for this operation is 3.2 pJ per operation. The total energy contribution of the multiplier units in this benchmark is 469 nJ, which is 34% of the total predicted energy. If the multiplier energy is scaled to 48% of its original value

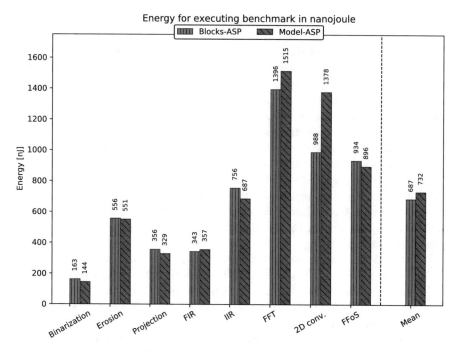

Fig. 6.9 Energy reported by ASIC synthesis versus energy reported by the energy model for Blocks with hard-wired connections

(the ratio between the multiplication and pass operations in terms of energy) the multiplier energy becomes 227 nJ. This brings the total energy for the benchmark to 1136 nJ and reduces the overestimation of energy to 15%.

The '2D convolution' benchmark shows that although in most cases it is sufficient to compute energy based on the data types used, there are cases in which the actual data values can have a significant impact on the estimated energy numbers. However, extending the model to take this into account would make it more complex and slower. If this level of accuracy is required it may be possible to have a compiler, generating code for the architecture, provide an estimation on the actual data range instead of just the data types. Especially in cases where values are constants, such as the coefficient value in the convolution benchmark, it should be possible to provide estimations on this. If the model is used for DSE such accurate absolute energy numbers are probably not required; the relative changes in energy between design iterations are more important for the cost functions used in such tools.

Besides providing an accurate measure of the total energy of an architecture instance, it is important to also estimate the individual contributions of parts of the design. This allows a designer (or automated tool) to gain insight into what part of the design is consuming the most energy. Figure 6.10 shows an energy breakdown, based on the post-synthesis power estimation provided by the

ASIC tool-flow (marked with 'B'). The energy breakdown is divided into several categories: the arbiter, the function units, the global memory (GM), the Instruction memories (IM), the Local memories (LM), the switch-boxes for both control and data networks (SWB), and other parts of the architecture that cannot be classified as any of these categories. The 'other' category includes communication bus interfaces, boot-loader, configuration scan-chain, and wrappers around memory blocks (for bypassing and automatic power-down). It also contains gates that are optimized out of the original module hierarchy by the synthesis tool and can no longer be linked anymore to a specific module. It is of course possible to force the synthesis tool to keep the hierarchy, but this prevents some optimizations from being applied. Since this part of the evaluation is based on the hard-wired Blocks architectures, there is no energy consumption for the 'SWB' category reported in this figure. Since this part of the evaluation is based on the hard-wired Blocks architectures, there is no energy number for the 'SWB' category in this figure. Figure 6.10 also shows the estimated energy distribution provided by the energy model (marked with 'M'). When comparing the estimated energy numbers and those provided by the ASIC tools it can be observed that the energy estimations for the function units and the global memory seem to be quite accurate. With the exception of the '2D convolution' and 'FFT' benchmarks where the multiplier energy, and thus the function unit energy, is overestimated.

The energy contribution for the arbiter seems to be underestimated, this may be due to the fact that for benchmarking the arbiter was evaluated separately from the

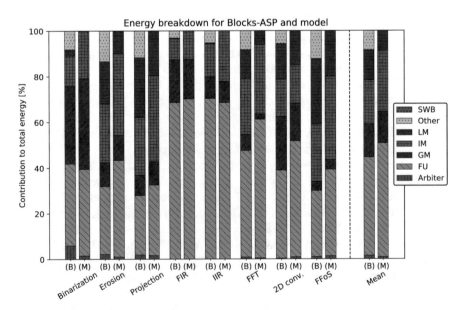

Fig. 6.10 Energy breakdown for the hard-wired architectures, according to the synthesis reports (B) and the model (M). Since the hard-wired architectures do not have a reconfigurable network, the contribution of switch-boxes to energy is zero

rest of the architecture (like all the architectural building blocks tested). However, in a real architecture the arbiter connects to LSUs that are spread throughout the design. It is possible that these location differences cause larger buffers to be selected in the data-path to meet timing requirements, leading to a higher energy consumption. The underestimation is the largest for the 'binarization' benchmark. In this benchmark the computation per byte of data is very low, leading to a high number of memory requests relative to the total number of cycles. Binarization is also the benchmark where the total underestimation of energy is the largest for all benchmarks. At least part of this is due to arbiter energy estimation. A possible solution for this is to scale the arbiter energy with the total area of the design.

For the local memories the energy contribution seems to be somewhat underestimated also. Part of this can be attributed to underestimation of leakage. This can be seen in, for example, the 'binarization' benchmark, where the local memories are not used. Where the synthesis report shows some energy consumption in the local memories, the model hardly shows any energy contribution for these memories. As the energy figures for the memories are directly based on those provided in the datasheet it is not clear what causes this difference. One of the causes might be that the data lines to the memory macros are currently not isolated, meaning that for some instructions these lines may be toggling even when no data is actually written to the memory. Both for real energy, as for estimated energy, it may be good to provide active isolation of these signals. However, this does increase the length of the path from LSU to local memory. If also applied to the global memory, then the isolation will be inserted in the critical path of the design, which should be avoided.

On the other hand, the energy for the instruction memories seems to be overestimated in the model. This is most likely caused by the fact that in the model every instruction is considered to be a change of value on the instruction memory interface. In reality, however, instructions do not always change between cycles. For example, an application may be designed to provide a loop body where the instruction is (almost) static. Although the instructions are still fetched from the instruction memories, the actual values on the instruction memory bus do not change.

Energy Estimation for Reconfigurable Instances

The reconfigurable architecture used to evaluate the benchmark kernels is the same for all kernels. For this reason, it contains the superset of function units required in the hard-wired architectures. The structure of the instantiated reconfigurable instance is the same as used in Chap. 5.

Estimating energy for the reconfigurable architecture instances includes estimating energy for unused function units and memories as well as the reconfigurable switch-box networks. As the switch-box energy makes up a reasonable fraction of the total energy consumption of the design, accurate estimation thereof is important. Figure 6.11 shows the reference and estimated energy numbers for reconfigurable Blocks architecture instances.

Similar to the hard-wired application specific designs, there is some under- and overestimation of energy. The '2D convolution' benchmark is still overestimated, but due to the higher total energy the overestimation is reduced to 29%. A new interesting case is the 'FFT' benchmark which is underestimated by approximately 19%. The other benchmarks have estimation errors between −20% and +14%. In general, the model, therefore, has a tendency to underestimate a little.

The reason that the error margins on reconfigurable architecture instances are higher can be explained by the fact that the energy consumption of the switch-boxes, and therefore the networks, is highly data dependent, much more so than for the function units and memories. This means energy differences due to data streams that are not behaving in a similar way as during micro-benchmarking are amplified in the reconfigurable networks. This is true for, for example, the 'FFT' benchmark. When comparing the breakdowns shown in Fig. 6.12 it can be observed that the switch-box energy is the cause of the underestimation. In fact, the 'FFT' benchmark has the highest switch-box energy for all evaluated benchmark kernels.

The cause of the underestimation lies in the data behaviour of the input data and intermediate data that is passed between function units. The 'FFT' benchmark operates on two's-complement numbers around zero, as the input signal is constructed from two added sine waves. The results are 8-bit values, stored as 32-bit numbers with sign extension to avoid having to do so at run-time. The sine waves cause

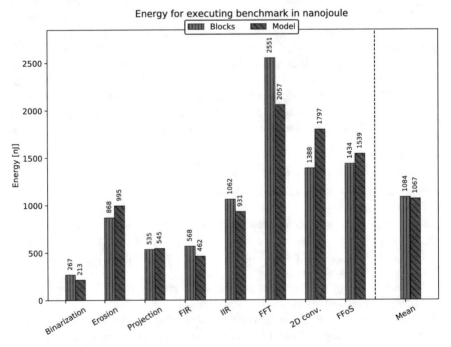

Fig. 6.11 Energy reported by ASIC synthesis versus energy reported by the energy model for Blocks with reconfigurable connections

the signal to move above and below the zero value. In two's-complement notation negative numbers have a sign-bit set to one and positive numbers a sign-bit set to zero. Due to the sign extension this causes the higher 24 bits to alternate between all ones or all zeros. This causes a higher number of toggling bits than the energy numbers in de energy model database are based on. The input vectors for the micro-benchmarks were constructed using independent randomly toggling bits. Therefore, an average of 16 toggling bits can be expected per cycle. While in the case of data for the 'FFT' benchmark this number is higher, leading to a higher switch-box energy in reality.

Figure 6.12 shows the energy distribution over the various parts of an architecture instance. Since reconfigurable architectures are compared, energy is consumed by the switch-boxes. Function unit energy is still very comparable to the estimates for the hard-wired architectures. Due to the presence of extra function units, which may be unused for some benchmarks, the area of the design is larger. When observing the actual and estimated energy fraction for the arbiter it shows that for reconfigurable architectures the underestimation is more significant than for the hard-wired architectures. This supports the hypothesis that the energy of the arbiter grows with the area of the design.

The local memory energy seems to be more underestimated compared to the hard-wired architectures. When the energy consumption reported in the post-synthesis report is compared for unused local memories, their energy numbers are about a factor two to three higher than the numbers based on the memory datasheets.

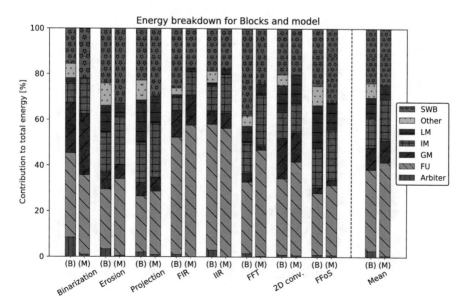

Fig. 6.12 Energy breakdown for the reconfigurable architectures, according to the synthesis reports (B) and the model (M)

This may indicate that there is more activity in the actual design than is assumed in the evaluation for the datasheet (e.g. no clock gating).

6.3.2 Area Model Evaluation

As there are less variables at play in the area model it may be assumed that estimating area should be more accurate. However, this is not necessarily the case, especially for the hard-wired architecture instances. Figure 6.13 shows the reported and estimated are numbers for the hard-wired architecture instances. The error of area estimation varies between −4% for 'erosion' and +4% for 'Binarization'. Other benchmarks with higher errors are 'FFoS' and 'Projection'.

For two of these benchmarks ('Erosion' and 'FFoS') the model underestimates the actual area. These benchmarks do not use all inputs and outputs of the available function units. Since the hard-wired architecture instances are essentially Application specific instruction-set processors (ASIPs), the unused inputs and outputs are unconnected or tied to ground for outputs and inputs, respectively. During synthesis the unconnected nets are detected and removed. For outputs this means that an entire output, including its multiplexer and buffer, can be removed. For inputs something

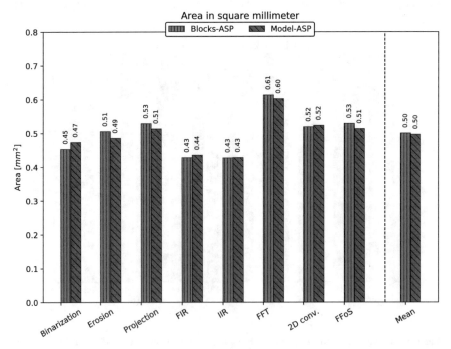

Fig. 6.13 Area reported by ASIC synthesis versus area reported by the area model for hard-wired connections

similar is performed, although in this case the constant values are not removed but
the logic they drive is replaced by another constant value. This iterative optimization
step can result in entire inputs and their multiplexers to be removed from the net-
list. Of course, this makes sense from a synthesis point of view but it makes area
estimation in these cases somewhat harder. It would be possible to extend the model
with an option that compensates for the optimization of one or more inputs or
outputs, but this would increase the model complexity. Currently, the area model
does not process any information regarding the connections between function units
and whether they are actually in use. However, it is possible to compute the number
of inputs and outputs that are in use.

The area breakdown in Fig. 6.14 shows that this is at least partially the
case but that there is more to the story. The area for the instruction memories
is overestimated. For these instruction memories the area numbers are directly
related to those in the datasheet. The reported area is directly obtained from those
reported by the synthesis tool, which in turn are obtained from the memory macro
description. For this reason it is not clear what causes this discrepancy. However,
the mean absolute percentage error over the benchmark set is 2.5%.

The area for the reconfigurable architecture instances can be accurately estimated
as all the inputs and outputs of the function units are connected to switch-boxes. The
paths through the switch-boxes are configured at run-time. As the select signals of
the internal multiplexers are therefore unknown during synthesis, it is not possible
for the synthesis tool to determine which paths are unused or unused. Due to this,

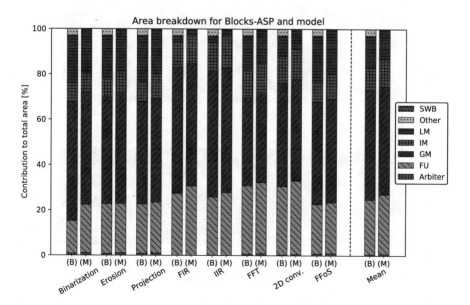

Fig. 6.14 Area breakdown for the hard-wired architectures, according to the synthesis reports (B)
and the model (M)

there are no paths or logic that will be removed, causing the area numbers in the model database to more accurately represent the area after synthesis. Since the benchmarks running on the reconfigurable architecture all use the same architecture description, the area is the same for all benchmarks.

The area of Blocks is 1.41 square millimetre and the model estimates an area of 1.38 square millimetre. Therefore, the mean absolute percentage error for the reconfigurable architecture instance is 2.1%. Figure 6.15 shows the area breakdown for the reconfigurable architecture. Similar to the hard-wired architecture instances, the area for the instruction memories is overestimated. There is also overestimation with regard to the switch-box area. This is because for the estimation of the area of a network a switch-box is used that has connections on all four sides (top, right, bottom, and left) as well as connections to the outputs and inputs of a function unit. This is not the case for all switch-boxes in the network, for example, a switch-box in the upper left corner of the grid structure will only have connections on the right and bottom as well as possibly connections to function unit inputs. This is something that could be improved upon the model, at the expense of some extra complexity. The reason for the slight underestimation, for the function units, can most likely be found in scaling of gates during synthesis to meet timing requirements. This could be

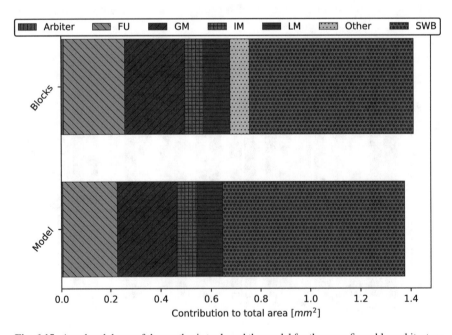

Fig. 6.15 Area breakdown of the synthesis tools and the model for the reconfigurable architecture

due to synthesis constraints. For example, the capacity on the outputs of a function unit can be different for the whole design compared to for a single synthesized unit. Alternatively, it can be that constraints on combinational paths are different, leading to different synthesis results. These differences are almost unavoidable when components are individually synthesized.

6.3.3 Model Calculation Time

Since execution time of the model is important when performing many DSE iterations, an analysis on the running time has been performed. Figure 6.16 shows model execution times for both the hard-wired (Model-ASP) and reconfigurable (Model) architectures. In both cases, the computation time for the energy and area results correlate with the number of cycles it takes to execute the benchmark application. This is caused by the fact that the model analyses an execution trace to obtain information for energy estimation.

The execution time for the model for a reconfigurable architecture takes longer to complete. This is mostly caused by estimating the energy for the switch-boxes. In order to do so, the place-and-route file specifying configured paths

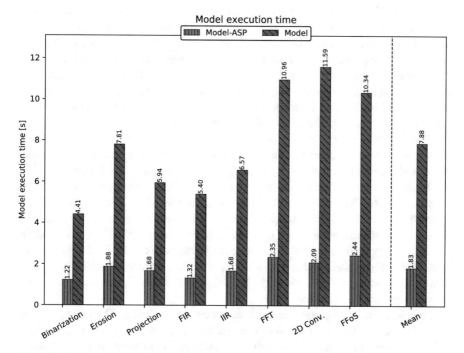

Fig. 6.16 Execution time of the model for the evaluated benchmarks. All measurements are performed with a single threaded application on an Intel Xeon E5-2620 v2

through the switch-boxes is analysed to extract the paths between function units. This calculation could also be performed when the architecture configuration is generated, moving this time away from the model and allowing some speed-up.

When compared to a full synthesis run, the model provides an average speed-up of close to three orders of magnitude. However, it is possible to make optimizations to the synthesis flow in some cases. For example, the design only has to be simulated if the instruction schedule changes. However, when function units are added, various aspects of the design change: the network, the arbiter, etc. In this case synthesis is required. However, it is possible to shorten the synthesis time by using pre-synthesized IP blocks for individual function components. After this synthesis the design has to be functionally simulated. However, in both cases, the model will still provide a significant speed-up.

Additionally, the model does not require access to ASIC technology libraries, which are often sensitive intellectual property. The energy and area model allows implementation of DSE tools for Blocks that provide fast energy, area, and performance trade-offs.

6.4   Conclusions

Reconfigurable architectures provide many options for configuration and application mapping. Finding a good implementation solution is, therefore, not a trivial task. Automated DSE can help a designer to find a good application mapping solution. In the case of CGRAs, DSE tools require both energy and area estimations to estimate whether a design will fit on a given amount of chip area and if it is more energy-efficient than another design. It is important that such estimations are both accurate and fast. Speed is important as a reduced estimation time will allow for more design iterations, possibly leading to better solutions.

The proposed energy model is calibrated on individual function units. These function units were stimulated on post-synthesis net-lists for all available instructions and a random sequence of input data with a predetermined data-width. As such, energy numbers are known for every data type supported by Blocks. This is a one-time effort. The numbers are used by the energy model, in combination with an instruction trace, to estimate the energy required to execute a given application. The instruction trace can be obtained by simulation or by a compiler. The results show that the energy model has a mean absolute percentage error of 10.7% for hard-wired and 15.5% for reconfigurable Blocks instances. The energy model demonstrates a maximum error of 39% and 29% for hardwired and reconfigurable instances, respectively, on the evaluated benchmark set.

Area is estimated by analysing the architecture description. The corresponding area for each of the building blocks in the design is looked up from a database of post-synthesis results. Summed up, the individual area numbers of subsystems provide the total design area. The area model has an error margin between -4% and $+4\%$ for hard-wired Blocks instances and -2% for reconfigurable instances.

Since the energy and area models do not require a full synthesis run of a design, the estimation time is significantly reduced. For reconfigurable architectures the model execution time is less than eight seconds, whereas a synthesis run takes over two hours to complete. Thus, the total execution time improvement is close to three orders of magnitude.

References

1. D. Liu, et al., Polyhedral model based mapping optimization of loop nests for CGRAs, in *Proceedings of the 50th Annual Design Automation Conference* (2013), pp. 1–8
2. Z. Zhao, et al., Optimizing the data placement and transformation for multi-bank CGRA computing system, in *2018 Design, Automation & Test in Europe Conference & Exhibition (DATE)* (IEEE, 2018), pp. 1087–1092
3. C. Liu, H. Kwok-Hay So, Automatic soft CGRA overlay customization for high-productivity nested loop acceleration on FPGAs, in *2015 IEEE 23rd Annual International Symposium on Field-Programmable Custom Computing Machines* (IEEE, 2015), pp. 101–101
4. D.L. Wolf, C. Spang, C. Hochberger, Towards purposeful design space exploration of heterogeneous CGRAs: Clock frequency estimation, in *2020 57th ACM/IEEE Design Automation Conference (DAC)* (IEEE, 2020), pp. 1–6
5. S.A. Chin, et al., CGRA-ME: A unified framework for CGRA modelling and exploration, in *2017 IEEE 28th International Conference on Application-specific Systems, Architectures and Processors (ASAP)* (2017), pp. 184–189. https://doi.org/10.1109/ASAP.2017.7995277
6. Y. Kim, R.N. Mahapatra, K. Choi, Design space exploration for efficient resource utilization in coarse-grained reconfigurable architecture. IEEE Trans. Very Large Scale Integr. (VLSI) Syst. **18**(10), 1471–1482 (2010). https://doi.org/10.1109/TVLSI.2009.2025280
7. S. Nalesh, et al., Energy aware synthesis of application kernels expressed in functional languages on a coarse grained composable reconfigurable array, in *2015 IEEE International Symposium on Nanoelectronic and Information Systems* (2015), pp. 7–12. https://doi.org/10.1109/iNIS.2015.39
8. A.S.B. Lopes, M.M. Pereira, A machine learning approach to accelerating DSE of reconfigurable accelerator systems, in *2020 33rd Symposium on Integrated Circuits and Systems Design (SBCCI)* (2020), pp. 1–6. https://doi.org/10.1109/SBCCI50935.2020.9189899
9. D. Chen, et al., High-level power estimation and low-power design space exploration for FPGAs, in *2007 Asia and South Pacific Design Automation Conference* (2007), pp. 529–534. https://doi.org/10.1109/ASPDAC.2007.358040
10. N. Dhanwada, et al., Efficient PVT independent abstraction of large IP blocks for hierarchical power analysis, in *2013 IEEE/ACM International Conference on Computer-Aided Design (ICCAD)* (IEEE, New York, NY, USA, 2013), pp. 458–465
11. A. Nocua, et al., A hybrid power estimation technique to improve IP power models quality, in *2016 IFIP/IEEE International Conference on Very Large Scale Integration (VLSI-SoC)* (IEEE, New York, NY, USA, 2016), pp. 1–6
12. R. Koenig, et al., KAHRISMA: A Novel Hypermorphic Reconfigurable-Instruction-Set Multi-grained-Array Architecture, in *2010 Design, Automation Test in Europe Conference Exhibition (DATE 2010)* (Mar. 2010), pp. 819–824. https://doi.org/10.1109/DATE.2010.5456939

13. K. Sankaralingam, et al., Trips: A polymorphous architecture for exploiting ILP, TLP, and DLP. ACM Trans. Architecture Code Optim. (2004)
14. B. Mei, et al., ADRES: An architecture with tightly coupled VLIW processor and coarse-grained reconfigurable matrix. *Field Programmable Logic and Application* (Springer, 2003)
15. E. Waingold, *Baring It All to Software: Raw Machines* (1997)
16. S. Thoziyoor, et al., *CACTI 5.1.*, Technical Report HPL-2008-20, HP Labs, 2008
17. T. Grosser, A. Groesslinger, C. Lengauer, *Polly - Performing Polyhedral Optimizations on a Low-Level Intermediate Representation* (2012)

Chapter 7
Case Study: The BrainSense Platform

All previous chapters in this book show how Blocks works and performs. Although Blocks can in principle work as a stand-alone device (currently without support for interrupts and exceptions), it is more likely that Blocks will operate as an accelerator in a SoC. Doing so requires integration of Blocks with one or more host processors and peripherals. On such a SoC it is common to find applications that exploit the heterogeneity of the platform. This means that parts of the application that perform best on one of the resources (processors, accelerators, etc.) are mapped to these resources. In this chapter a case study of such a system is performed. The system used as an example is a Brain-computer interface (BCI). These systems are able to provide computer input based on detected brain signals. One application thereof is brain controlled typing.

The remainder of this chapter is organized as follows. Section 7.1 provides an introduction in BCI systems and the BrainSense concept. Section 7.2 introduces the BrainSense platform, and its design choices, at a system level. Section 7.3 goes into further detail by providing analysis of the algorithm and introducing the Blocks-based accelerator. Section 7.4 provides energy estimations, based on the model from Chap. 6, for the BrainSense platform. Finally, Sect. 7.5 concludes this chapter.

7.1 Background

The BrainSense platform is not the first to focus at BCI. However, the algorithm and underlying method for doing so is significantly different than existing work. This section describes some of the existing work in BCI platforms, as well as a short introduction into the underlying concept.

© The Author(s), under exclusive license to Springer Nature Switzerland AG 2022   181
M. Wijtvliet et al., *Blocks, Towards Energy-efficient, Coarse-grained Reconfigurable
Architectures*, https://doi.org/10.1007/978-3-030-79774-4_7

7.1.1   Existing BCI Platforms

The concept of using signals measured from the brain has been explored before. An example thereof is the BCI2000 [1] which uses Electroencephalography (EEG) signals, picked up by electrodes on the skull, as a way to interface with devices. The BCI2000 system interprets brain signals directly, meaning that the user has to 'imagine' a specific task or movement. This pre-trained signal then gets recognized and a certain interface task is activated. The BCI2000 system is running on a general purpose computer and not aimed at wearable implementations. Recognizing these brain signals is considerably more complicated and error prone than the method that the proposed BrainSense system uses, which is based on gold codes projected on the screen of the target device that the user is interacting with [2, 3]. This also means that interacting is far more natural and requires no specific training for the user.

Other biomedical signal processors have been proposed in the past which aim to improve energy efficiency. These gains can be attained in several ways, in [4], for example, the authors propose a biomedical signal processing platform that consists of a microprocessor and some fixed accelerators that can be reached over a bus. In this design the algorithms that match well to the available accelerators will achieve a speed-up and an energy efficiency gain, for other applications the gains will be minimal. Processors can also be tuned for a specific domain. An example thereof is [5], which is specialized in feature classification and classification based on support vector networks. The architecture in this example is, in fact, so specialized that it will be challenging to run any other application on it. In contrast, the proposed BrainSense architecture aims to avoid too extensive specialization by using a CGRA as a reconfigurable accelerator.

CGRAs have been proposed for EEG related applications, examples thereof are [6] and [7]. These CGRAs support typical EEG feature extraction algorithms such as FIR and IIR filters as well as FFT. The structure of these architectures provides a data-path aimed at streaming data. The operations are statically defined and the CGRA is effectively used as a systolic array. In contrast, the Blocks CGRA used in BrainSense does support branching and, with this, program flow control. Instead of implementing systolic arrays it can efficiently implement a run-time reconfigurable application specific processor. Blocks uses its unique reconfigurable instruction decoder network which allows construction of SIMD structures, allowing efficient implementation of data parallelism.

As discussed in [3], BBEP-based BCIs have been predominantly studied in the visual domain using the so-called Broad-band visually evoked potentials (BBVEPs), or Code-modulated visually evoked potentialss (cVEPs). The field was initiated in 1984 and first tested with an ALS patient who was able to spell about 10 to 12 words per minute with a BBVEP-based speller and intracranial recordings [2, 8]. Since then, BBVEPs have proven to be successfully decodable from EEG and to enable fast and reliable communication [9–13]. The BCI based on this technique (known as re-convolution) achieves a spelling speed of one letter per second, using small headsets with only eight electrodes. This technology has a wide range of other applications that can be explored in the future (such as virtual reality, sleep assistance, home automation, and entertainment).

7.1.2 The BrainSense BCI Concept

The algorithm used in the BrainSense platform is developed by the Donders Institute for Brain, Cognition and Behaviour at the Radboud University [2, 3]. Data is obtained using electrodes placed on the head of the user. These electrodes make contact with the skin and pick up electronics potentials generated by the brain. Most EEG signals work by interpreting signals generated by the brain itself. For example, the user can navigate a menu by thinking about a certain action, e.g. pushing an object 'forward' to press a button in a menu. Classifying brain signals to provide a reliable and fast way of computer interaction is very hard. For this reason, the algorithm used on the BrainSense platform uses external stimuli to influence the brain signal of the user. This is achieved by presenting visual cues on a screen. The visual cues consist of 'flickering' patterns. The presented patterns are modulated as gold codes. This algorithm functions by recognizing gold codes in the measured electrode signals. The user can control the pattern that is measured at the electrodes by focusing on one specific pattern. Figure 7.1 shows a user typing using a BCI.

To enable a user to type using the algorithm, three stages exit. In the first stage stimulations are applied and the response at the electrodes is measured, this is called the training stage. Based on these responses the second stage can perform calibration. During the calibration stage the responses to short and long pulses in the stimuli are determined. These responses can be used to construct expected responses in the electrode signals to applied stimuli. By doing so, a predicted response can be constructed for each gold code that will be shown on the screen. Using correlation it can be determined which predicted response is closest to the actual response in the

Fig. 7.1 A user typing using a BCI. Each of the grey squares on the screen blinks in a unique pattern and represents a specific letter on a keyboard

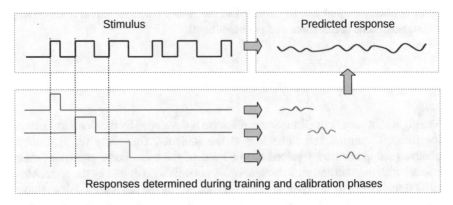

Fig. 7.2 Algorithm flow for the prediction phase, this is performed for every gold code displayed on the screen. A stimulus is generated to invoke a signal at the electrodes on the back of the head of the user. Responses obtained during the training phase are used to compute a predicted brain response. This figure is adapted from [3]

third stage, the testing stage. If the confidence on the detected signal is sufficiently high it will classify as a detected letter (or a pressed button) and used as a 'typed' letter. Figure 7.2 shows how the algorithm to predict a brain response works for one transmitted code. This is performed for every code shown on the screen.

7.2 System Level Design

In almost every use-case for BCI devices wearability is an important aspect. Even in cases where freedom of movement is not critical, a small wireless device will be much less obtrusive than a traditional EEG-like headset that needs to be connected (wired) to a computer. New BCI algorithms perform accurate detection with only a few electrodes on the back of the head of the user, allowing for a physically very small device. This places constraints on the size, and therefore capacity, of the battery contained in the headset. This, in turn, leads to a requirement for high energy efficiency in computation and signal acquisition.

The BCI headset needs to communicate with the target device, typically a tablet or computer, via a wireless link. Wireless communication is expensive in terms of energy and, therefore, needs to be limited as much as possible. This means that the typical approach of wireless transmission of raw electrode measurements to the target device for further processing is not desirable. Instead, the electrode signals should be preprocessed as much as possible on the headset itself, reducing the amount of data to only the relevant information. When the data can be fully processed on the BCI headset only the detected interaction needs to be communicated to the target device (e.g. the letter 'A' has been typed). It is, therefore, required that the processing platform will be able to detect a user interaction sufficiently fast,

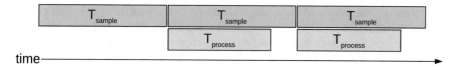

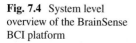

Fig. 7.3 Pipelined acquisition of electrode measurements and processing. The sample rate of the current BrainSense algorithm is 360 Hz, allowing for 2.8 ms of processing time. Eight channels are samples in parallel

Fig. 7.4 System level overview of the BrainSense BCI platform

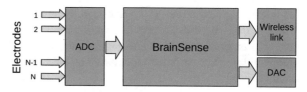

meaning that the platform should take less time to process a batch of samples than the time required to obtain these samples. This allows an implementation where samples can be recorded and processed as batches in a pipelines fashion, as shown in Fig. 7.3. Doing so puts constraints on the processing time and thus requires a certain amount of processing power.

The remainder of this section will introduce subsystems of the BrainSense platform. Section 7.2.1 introduces the peripherals that are required to interface with the analogue interface as well as the wireless communication module. Section 7.2.2 introduces the BrainSense processor. Section 7.2.3 concludes this section by introducing the memory hierarchy and interconnect subsystem.

7.2.1 Communication and Peripherals

The input to the BrainSense platform is the set of measured potentials at several electrodes placed on the head of the user. The algorithm used for BrainSense uses eight electrodes. Seven of these electrodes are used as measurement electrodes, the eighth is used as a reference electrode. To obtain the electrode values in a digital form, an analogue front-end and analogue-to-digital conversion are required. Furthermore, it is envisioned that active noise cancellation will be performed on the shielding of the electrode wires. This requires the system to control the potential on the shielding. Finally, an interface needs to be present to transmit the detected input from the BrainSense headset to a target device. These requirements lead to a system overview of the proposed BrainSense platform, as shown in Fig. 7.4.

As can be observed, the electrode potentials are measured using an Analog-to-digital converter (ADC) with multiple inputs. For the BrainSense platform described in this work, the number of electrode inputs is eight. Additionally there is a Digital-to-analog converter (DAC). The DAC is used to perform active shielding of the electrode wires. It is controlled by the BrainSense platform to cancel out external interference. The details of the analogue front-end are out of the scope of this book. The wireless link is used to communicate by the target device and is managed by the BrainSense platform via a serial protocol. For ease of integration a Bluetooth Low-Energy (BLE) device is used.

The processing platform for BrainSense must be able to control the ADC for digitizing the electrode potentials, the DAC for providing active electrode shielding, and a wireless link to communicate data. The ADC envisioned for the BrainSense platform communicates via the serial peripheral interface (SPI). The DAC uses an inter-integrated circuit (I^2C) interface. Finally, the wireless link communicates via a UART. In order to provide some way to directly control status indicators and read external buttons, some general purpose IO (GPIO) pins are added to the peripheral set. The peripherals are based on the Pulpino [14] platform and allow I/O multiplexing.

7.2.2 The BrainSense Processor

BrainSense requires a compute platform to process the recorded electrode signals. One of the most popular microprocessors for digital signal processing on wearable (medical) devices is the ARM Cortex-M4f, a 32-bit microprocessor with hardware floating point support. The aim of the BrainSense platform is to provide an energy-efficient platform for signal processing in BCI applications. Since the Cortex-M4f is such a popular processor for this application domain, many applications have been written with this processor in mind. The BrainSense platform utilizes this processor and extends it with a CGRA based accelerator for BCI, as shown in Fig. 7.5. This allows migration to BrainSense without requiring drastic application changes, the designer can move the desired kernels to the Blocks CGRA one by one.

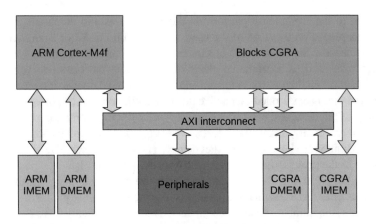

Fig. 7.5 Overview of the BrainSense BCI processor architecture

Blocks only supports fixed point arithmetic out of energy efficiency considerations; however, parts of the application that cannot be converted to fixed point (e.g. due to convergence problems) can be mapped to the microprocessor.

For BrainSense, Blocks will be configured with a set of functional units tuned to the main BCI workload for inference. Although there is a significant amount of processing going into the learning stage of the algorithm, this step is currently only performed once. In the future the learning stage will be performed regularly during use. The main workload of the algorithm is based on filtering, essentially convolution in various forms. The microprocessor orchestrates the execution of the BCI algorithm. It configures the reconfigurable accelerator to perform a certain kernel and prepares the data it needs for execution and manages execution.

To enable communication, the microprocessor and Blocks are connected over an AXI interconnect and can both access the peripherals. This leads to a flexible platform, shown in Fig. 7.5, that is suitable for the current BCI algorithm, yet can be updated when the algorithm changes.

7.2.3 Memory and Interconnect

Operation of the Cortex-M4f microprocessor requires an instruction memory and a data memory, connected to their respective buses on the processor. The instruction and data memories for the Cortex-M4f are privately connected, as shown in Fig. 7.5, as there is no need for the CGRA to be able to read from, or write to, these memories. The Cortex-M4f can access the BrainSense peripherals over the AXI interconnect. The microprocessor can control execution via the control interface of the CGRA over the AXI interconnect. It allows starting, stopping, and restarting the loaded kernel. Additionally, a status register can be used to monitor the execution.

The Blocks accelerator has its own instruction memory. This memory can also be accessed by the host processor to provide executable binaries to Blocks. Upon commands of the host processor, Blocks loads the relevant binary, configures the interconnect networks, and prepares the local instruction memories. Data is provided to Blocks via the data memory. This memory can be accessed, over the interconnect, by both the host processor and Blocks. Therefore, this memory can be used for data exchange. However, care should be taken as there is no system to prevent overwriting data by one device that may be in use by the other. The memory sizes for the ARM Cortex-M4f data and instruction memories are chosen similar to what can be found on most microprocessor systems using this processor, namely 64 kilobyte in total (two memories of 32 kilobyte each). This allows the BrainSense platform to provide an easy target for applications moving from a single microprocessor system to a system-on-chip such as BrainSense.

Both the Cortex-M4f and Blocks can act as a master on the interconnect; arbitration is handled by the interconnect. Since the peripherals are connected to the interconnect, both Blocks and the microprocessor can control the peripherals.

Therefore, control of the ADC, DAC, and wireless link can be under control of both devices. This allows maximum flexibility at run-time.

7.3 Algorithm Analysis and Acceleration

Future versions of the BrainSense algorithm perform online training. The envisioned training update rate is about once per second. To do so, the algorithm performs four stages: (1) preprocessing, (2) Updating summary and statistics (USS), (3) Canonical Correlation Analysis (CCA), and (4) classification.

Multiple samples of electrode data are grouped in batches, called epochs, and are used by the algorithm to compute a single result. The computation required for the first epoch involves all four processing steps as it performs calibration for each spelled letter. A typical typing rate is approximately one second, requiring 29 epochs to be processed. For the remaining epochs only the preprocessing and classification steps are required, this leads to a processing scheme as shown in Fig. 7.6.

Figure 7.7 shows the processing time distribution for each of these steps of the algorithm. The execution times in this section are taken from [15]. The tasks 'update statistics' and 'correlation analysis' implement the training stage described in Fig. 7.6.

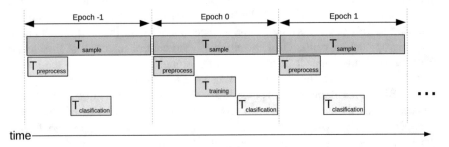

Fig. 7.6 Processing required for one epoch. The first epoch for each spelled letter requires training to be performed, subsequent epochs only require filtering and classification

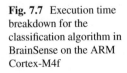

Fig. 7.7 Execution time breakdown for the classification algorithm in BrainSense on the ARM Cortex-M4f

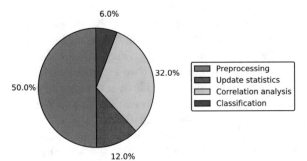

Fig. 7.8 The Blocks instance used for the BrainSense platform

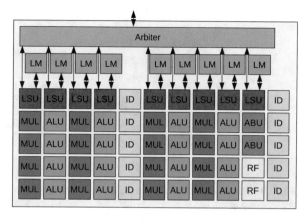

As can be observed, the main time consuming task on the microprocessor is the preprocessing. Approximately 80% of the time in preprocessing is spent in filtering, the remaining time is spent in sub-sampling and converting data streams into epochs. After preprocessing, the largest amount of time is spent in CCA. Within CCA, a little over 50% of the time is spent in performing matrix multiplication. The other half is taken up by single value decomposition (SVD). The largest gains can be obtained by speeding-up the preprocessing steps and can be achieved by optimizing the preprocessing step and CCA. For this reason, these parts of the algorithm are investigated to be ported to the Blocks framework.

All performance evaluations for Blocks are performed on the architecture instance shown in shown in Fig. 7.8.[1] This architecture is used for matrix multiplication as well as the IIR filter and classification. This architecture instance implements a 4×4 tiled matrix multiplication, a size that matches well with the matrix sizes used in the CCA algorithm.

In [15] speed-ups are reported of $4 \times$, $9.9 \times$, $21.2 \times$ for matrix multiplication of sizes 8×8, 16×16, and $32 \times 32 \times$, respectively, when accelerated on Blocks. These implementations use approximately 36 cycles to compute one output matrix element for the small 8×8 matrices and seem to converge to around 26 cycles per matrix element for the larger matrices.

Above numbers are used to estimate the number of processing cycles required to compute the matrix multiplications in the CCA algorithm. Matrix multiplications of several sizes are performed within the CCA algorithm; these sizes are 8×8, 38×38, 8×38, 8×1, and 38×1. The number of cycles performed in matrix multiplication can be estimated to be in to order of 50,000 cycles. Compared to over 400,000 cycles for matrix multiplication on the ARM Cortex-M4f the CGRA will provide a speed-up in matrix multiplication of approximately $8 \times$, leading to a total speed-up of the CCA algorithm of $1.8 \times$.

---

[1]The CGRA instance for matrix multiplication, IIR, and classification on the BrainSense platform is developed by Joris Witteman as part of his Master thesis assignment.

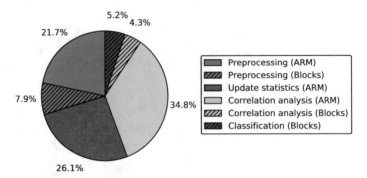

Fig. 7.9 Execution time breakdown for the BrainSense algorithm with acceleration on Blocks for spelling one letter. The shaded parts mark the execution fractions of the Blocks accelerator

Similarly, a speed-up is reported of 14× for an IIR filter. The IIR filter is used for filtering in the preprocessing step and takes about 80% of the computation time of this step in the algorithm. With the reported speed-up, the total performance gain for the preprocessing is over 3.9×. The last step of the algorithm is classification, this step can be ported entirely to Blocks and results in speed-up of 2.5×.

The first epoch is accelerated by over 1.5× by porting the filtering in the preprocessing, the matrix multiplication, and the classification to Blocks. The subsequent epochs provide a speed-up of over 3.7× since these only require preprocessing and classification, the parts of the algorithm where Blocks provides the highest gains. For a full spelling sequence this provides a performance improvement of 2.2×. This performance gain can be used to reduce the processing duty-cycle, and switch the processing off when not required. Or, use the extra time available to reduce the frequency and voltage of the system in order to improve energy efficiency. Figure 7.9 shows the execution time breakdown for the BrainSense algorithm with acceleration on Blocks included.

It can be observed that with acceleration the computation bottleneck moved from the preprocessing stage to the correlation analysis. Blocks only accelerates about half of the computation performed in this kernel. Further optimizations should be investigated by providing and efficient mapping of the SVD algorithm to Blocks, which is currently still performed mostly on the ARM. The SVD algorithm takes approximately 89% of the processing time for CCA after matrix multiplication has been accelerated by Blocks. By performing the remaining steps (sub-sampling and grouping samples into epochs) in the preprocessing step on Blocks another reduction of processing time can be achieved. The same holds for updating the statistics in the USS step. In order to do so, the covariance has to be computed. With a smart selection of the number of samples it should be possible to avoid division, which is currently unavailable in Blocks, and use arithmetic shifts instead. However, extending the Blocks multiplication unit with division operations may lead to a more generically applicable approach. If the remaining parts that are currently processed on the ARM processor can be mapped to Blocks with similar speed-ups as for the accelerated kernels, a total speed-up in the order of 8× should be achievable.

7.4 Evaluating BrainSense Energy Consumption

To estimate the energy gains for the BrainSense application, the energy model for Blocks is used to predict the energy numbers for the preprocessing, CCA, and classification stages. To do so, benchmark kernels for an IIR filter, matrix multiplication, and classifier were simulated in order to obtain an instruction trace. Subsequently the energy model was used to estimate the energy per cycle for each of these benchmarks. Based on the number of cycles determined for these kernels in Sect. 7.3 the energy for one complete detection of a letter can be computed.

The energy model for Blocks cannot be used to predict energy numbers for the ARM processor. Instead the energy numbers and cycle counts for the processing stages reported in [15] were used to obtain energy per cycle. By multiplying the energy per operation numbers with the cycle counts obtained after acceleration, the energy for the ARM processor can be estimated. The energy for USS is not reported, therefore the same energy per cycle as for matrix multiplication is used for this processing stage. This can of course introduce some error in the energy estimations, but the energy per operation for each of the processing stages on the ARM processor is relatively similar. These numbers include accesses to the instruction and data memories.

Figure 7.10 presents the energy numbers, for one spelled letter, for each of the processing stages in the BrainSense algorithm. Updating the statistics is currently not accelerated and, therefore, uses the same amount of energy in the original and accelerated versions. The other three stages of the algorithm benefit from acceleration in terms of energy. It can be observed that the shaded areas, which mark energy used by the Blocks accelerator, are relatively small. This indicates that the Blocks accelerator performs a good job in accelerating these parts of the algorithm in an energy-efficient way.

The energy gains are largest for the preprocessing stage. In this stage 80% of the cycles performed on the ARM processor could be offloaded to Blocks. The energy gain is roughly equal to the speed-up of this part of the algorithm, this can be explained by the reduced number of accesses to memory as the intermediate results are kept in the Blocks pipeline but do not fit in the register file of the ARM. The correlation analysis benefits less from acceleration, mostly because the SVD is not mapped to Blocks and still consumes a significant amount of cycles (and energy) on the ARM.

For all accelerated parts of the algorithm it is very clear how Amdahl's law comes into play. Blocks accelerates parts of the algorithms significantly, but this makes the other parts dominate the energy and performance metrics. In terms of energy, the ARM processor uses 87% of the total energy used by the accelerated platform. If the kernels running on the ARM cannot be accelerated, further energy gains will be minimal. However, as described in Sect. 7.3 it is likely that the preprocessing stages and statistics update can be accelerated. If these scale similar to the currently accelerated kernels the energy gains would likely increase from $2.3\times$ to around $5\times$. At this point the SVD will completely dominate the energy usage (70%) of

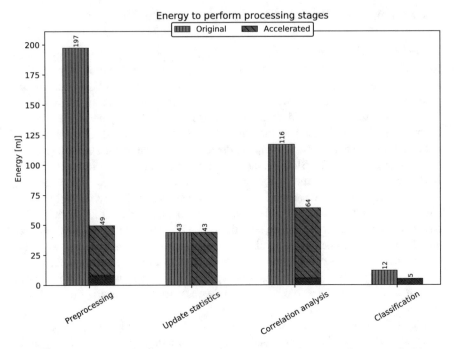

Fig. 7.10 Energy required to perform each of the four processing stages for one spelled letter. The densely shaded areas indicate energy used by the Blocks accelerator, the lightly shaded areas indicate energy used by the ARM Cortex-M4f processor

the system as it will be the only part of the algorithm still running on the ARM processor. For BrainSense it makes sense to further investigate how the SVD can efficiently be mapped to Blocks, and whether any hardware changes would be required to do so.

7.5   Conclusions

Blocks can operate either stand-alone or as part of an integrated system-on-chip. This chapter shows how Blocks can be used as an accelerator for a brain–computer interface (BCI). The application used on this platform includes four processing stages to obtain a spelled letter from brain signals. The brain signals are influenced with external blinking sequences, these sequences are unique for each letter. By focusing their attention on one of the letters users can influence which signal is detected strongest by electrodes on the back of the head.

Based on an evaluation of the algorithms, candidates for acceleration on Blocks are selected. The performance improvement for parts of the algorithm reach increases as high as 14×. However, there are parts of the algorithm that are still to be optimized for execution on an accelerator. This restricts the total speed-up to 2.2×. These performance estimates are used to obtain estimated energy numbers for Blocks (as part of the SoC) and show an energy improvement of 2.3×.

However, predictions estimate that a total speed-up in the order of 8× should be achievable. This indicates that energy improvements around 5× should be achievable but that further gains are still restricted by the part of the algorithm running at the host processor.

References

1. G. Schalk et al., BCI2000: a general-purpose brain-computer interface (BCI) system. IEEE Trans. Biomed. Eng. **51**(6), 1034–1043 (2004)
2. J. Thielen et al., Broad-band visually evoked potentials: re(con)volution in brain-computer interfacing. PLoS One **10**(7), e0133797 (2015)
3. J. Thielen et al., Re (con) volution: accurate response prediction for broad-band evoked potentials-based brain computer interfaces, in *Brain-Computer Interface Research* (Springer, New York, 2017), pp. 35–42
4. J. Kwong, A.P. Chandrakasan, An energy-efficient biomedical signal processing platform. IEEE J. Solid-State Circ **46**(7), 1742–1753 (2011). ISSN: 0018-9200. https://doi.org/10.1109/JSSC.2011.2144450
5. J. Yoo et al., An 8-channel scalable EEG acquisition SoC with patient-specific seizure classification and recording processor. IEEE J. Solid-State Circ. **48**(1), 214–228 (2013). ISSN: 0018-9200. https://doi.org/10.1109/JSSC.2012.2221220
6. H.H. Avelar, J.C. Ferreira, Design and evaluation of a low power CGRA accelerator for biomedical signal processing, in *2018 21st Euromi-cro Conference on Digital System Design (DSD)*, August 2018, pp. 488–491. https://doi.org/10.1109/DSD.2018.00087
7. J. Lopes, D. Sousa, J.C. Ferreira, Evaluation of CGRA architecture for real-time processing of biological signals on wearable devices, in *2017 International Conference on ReConFigurable Computing and FPGAs (Re- ConFig)*, December 2017, pp. 1–7. https://doi.org/10.1109/RECONFIG.2017.8279789
8. E. Donchin, L.A. Farwell, Talking off the top of your head: toward a mental prosthesis utilizing event related brain potentials, in *Electroencephalography and Clinical Neurophysiology* (1988) 10.1016/0013-4694(88)90149-6
9. S. Gao et al., Visual and auditory brain-computer interfaces. IEEE Trans. Biomed. Eng. **61**(5), 1436–1447 (2014). https://doi.org/10.1109/TBME.2014.2300164
10. B. Blankertz, M.S. Treder, Research covert attention and visual speller design in an ERP-based brain computer interface. Behavioral & Brain Functions, in *BioMed Central* (2010)
11. I. Volosyak, SSVEP-based Bremen-BCI interface-boosting information transfer rates. J. Neural Eng. **8**(3), 036020 (2011). https://doi.org/10.1088/1741-2560/8/3/036020
12. G. Bin et al., VEP-based brain-computer interfaces: time, frequency and code modulations. IEEE Comput. Intell. Mag. **4**, 22–26 (2009). https://doi.org/10.1109/MCI.2009.934562
13. E.E. Sutter, The visual evoked response as a communication channel, in *Proceedings of the IEEE Symposium on Biosensors* (1984)
14. PULP, PULP platform https://www.pulp.platform.org/. Accessed 2019 November 28
15. J. Witteman, Energy efficient brain controlled typing in 40 nm CMOS (Master Thesis report). Technische Universiteit Eindhoven, 2019

Chapter 8
Conclusions and Future Work

This chapter reviews the work presented in this book. Furthermore, future research paths and extensions to this book are presented.

8.1 Conclusions

Embedded processors can be found in most electric devices nowadays. These processors range from very simple, merely there to control the timer on a microwave, to very advanced multi-processor platforms like those found in smart-phones. For battery powered devices, these processors are often highly optimized towards a specifically targeted application domain. If production volume is large enough, then dedicated chips can even be made. Specialization is performed to achieve high energy efficiency on such devices. However, specializing processors at design time leads to a reduction in flexibility. For some cases this is no problem; the processor will never be required to perform another function. But other devices may be required to perform multiple tasks. The processor on a smart-phone for example needs to take care of communication, user interfacing, and digital signal processing. This requires processors to be flexible.

Flexibility usually does not combine well with a requirement for energy efficiency as the increasing number of possible configurations results in a more complex system and a higher power draw. On the other hand, a flexible system that can enable a more efficient application mapping will lead to higher performance and a reduction in energy. Field programmable gate arrays (FPGA) are very popular for devices where reconfigurability is of high importance. However, their bit-level reconfigurability leads to a very high energy overhead. Coarse-grained reconfigurable architectures (CGRAs) are more suitable to the digital signal processing found on most devices. Most CGRAs implement either a quite static systolic array or provide a grid of lightweight processors that can communicate over a network.

M. Wijtvliet et al., *Blocks, Towards Energy-efficient, Coarse-grained Reconfigurable Architectures*, https://doi.org/10.1007/978-3-030-79774-4_8

In Chap. 1 Blocks is proposed, the goal of Blocks is to allow instantiation of application specific processors on a reconfigurable fabric. This enables the use of both data-level and instruction-level parallelism as well as combinations thereof, leading to energy efficiencies normally only found in fixed processor layouts, such as SIMD and VLIW processors.

In Chap. 2, a definition of a course grained reconfigurable CGRA is provided. In this book, an architecture is considered a CGRA if it provides a reconfiguration granularity at the level of a fixed function unit or larger, and if its temporal reconfiguration granularity is at the region/loop-nest level or above. The first requirements rule out bit-level reconfigurable architectures such as FPGAs. The second rules out processors that reconfigure on a cycle-based temporal level. Such processors can be typical general purpose processors but also many-core systems-on-chip. To validate this definition, a total of 40 CGRA architectures that have been presented in the past are investigated. The properties of each architecture are summarized and classified on structure (network, function units, and memory hierarchy), control (scheduling, reconfiguration methods, and supported parallelism types), integration (how the CGRAs interact with a host processor, if any), and tool support (what compilers, mapping, and design space exploration tools are available). Based on these properties, an overview is presented on how these CGRAs fit within the CGRA definition. The chapter concludes with observations on the past CGRAs. Such observations included the preference for two-dimensional mesh networks as well as indications that most architectures in the past have aimed at high-performance processing rather than energy efficiency. The chapter concludes with five research guidelines based on these observations.

Chapter 3 describes the main concept of Blocks: separation between function unit control and the data-path. This separation is implemented with two independently configurable networks, a data and a control network. The control network allows instruction decoders to be connected to one or more function units. Subsequently, these units are under the control of the instruction decoder. This allows flexible formation of vector lanes, multiple function units controlled by a single instruction decoder. This contrasts to typical CGRAs that have either a global context (static or dynamic) or local decoding per function unit. A distributed program counter allows the instruction decoder to operate in lock-step, thus forming VLIW processors. Doing so allows construction of application matched processors. The expectation is that support for multiple types of parallelism combined with specialized, run-time configurable, data-paths leads to high energy gains. Furthermore, the available function units (ALU, ABU, LSU, RF, IU) are introduced and their design choices explained. Finally, thoughts on how multi-processor structures and approximate computing can be implemented on Blocks are described.

The tool-flow of Blocks is described in Chapter 4. Blocks uses a two-stage approach to generate a synthesizable reconfigurable hardware design. First, a virtual architecture is specified by the user. It is specified in an XML format, specifying which function units are present and how they are connected to each other. This description can be synthesized, resulting in an application-specific processor.

For this processor applications can be written in a parallel assembly language, called PASM. The virtual architectures allow relatively fast RTL-level simulations and more convenient debugging than the fully reconfigurable architectures. Once the virtual architecture and the application running on it are functioning, the virtual architecture can be mapped to a physical architecture, this is the second step. For this step, a physical architecture must be specified. It describes the available function units and their location in the reconfigurable networks. Properties of the network and the function units are design-time parameters. Mapping virtual architectures onto physical architectures is an automated process. The chapter continues by detailing the design-time parameters that can be adjusted. Furthermore, the steps required to produce an executable binary from an architecture description and PASM is described. To do so, PASM is assembled into machine instructions and an architecture description is translated into a bit-stream. The bit-stream specifies the network and function unit configurations. Finally, the bit-stream and machine code are merged into an executable binary.

Chapter 5 evaluates the performance, area, and energy efficiency for Blocks. The results show that Blocks, due to separation of data-path and control-path, significantly reduces reconfiguration overhead even when compared to an already optimized, but more traditional, CGRA. All evaluations are based on post-place-and-route results of fully operational RTL implementations of all architectures. Compared to a traditional CGRA, Blocks reduces reconfiguration energy overhead between 46% and 76% (average 60%), depending on the benchmark, without performance penalty. The system level energy reduction is between 9% and 29% (average 22%). This shows that separation of control and data for energy-efficient CGRAs makes sense. For many applications, especially when data-level parallelism is available, Blocks will achieve better energy efficiency. Furthermore, Blocks enables area reduction, compared to traditional CGRAs, for applications where data-level parallelism can be exploited, due to the reduction of instruction decoders and associated instruction memories. Blocks is also compared against fixed archi-tectures, an 8-lane SIMD with control processor and an 8-issue slot VLIW. The geometric mean of the area efficiency (performance per area) improvement of Blocks is 1.2x and 1.5x over the (fixed) SIMD and VLIW, respectively. Despite reconfiguration overhead, energy consumption is 2.1x and 1.8x lower for Blocks. Furthermore, Blocks energy is over 8x lower compared to an ARM Cortex-M0, while achieving a speed-up of 68.9x. These results show that although there is a price to pay for flexibility, it might be lower than expected. To the best of our knowledge, this is the first CGRA architecture comparison based on post-place-and-route layouts of full processor designs on a 40nm commercial library. When performance per area is considered, Blocks even outperforms the fixed architectures and reference CGRA. In case of the fixed architectures, this is achieved by higher performance, while in case of the reference CGRA, this is achieved by an area reduction.

To allow fast design space exploration, an energy and area model is required. Chapter 6 details how the model for Blocks is developed. The model is based on profiling individual function units using post-synthesis designs. To do so, test-benches are constructed that allow stimulating the function units with input vectors. The input vectors control the data supplied to the function units as well as the instructions they perform. This allows energy analysis for each instruction and for a given number of toggling input bits, resulting in an energy model database for each function unit. Area is obtained directly from the reported area numbers for the synthesized function units. The energy model uses an instruction trace to estimate activity in the switch-boxes and function units. The activity is based on which instruction is active for a function unit in a cycle of operation, the energy for this is then extracted from the model database. Estimating area is easier as it does not depend on the activity of the architecture. This is also visible in the evaluation of the models. The area model has an error between −4% and +4% for hard-wired architectures and an error of −2% for reconfigurable architectures. The energy model is somewhat less accurate with an average overestimation of 7% and an underestimation of 2% for hard-wired and reconfigure architectures, respectively. When the execution time of the model is taken into account, these are very acceptable results. Whereas synthesis takes a least two hours to complete, the average execution time of the model is less than eight seconds, an improvement of close to three orders of magnitude.

Chapter 7 presents a case study for a system-on-chip that uses Blocks to increase energy efficiency for processing an algorithm. The system-on-chip used for this case study is a brain–computer interface, a system that allows users to type by using brain signals. The algorithm used for this is developed by the Donders Institute for Brain, Cognition and Behaviour. It includes four processing stages to obtain a spelled letter from brain signals. The brain signals are influenced by external blinking sequences; these sequences are unique for each letter. By focusing on one of the letters users 'select' which signal is detected strongest by electrodes on the back of the head. Based on an algorithmic evaluation, candidates for acceleration on Blocks are selected. Selected parts of the algorithm achieve speed-ups in the order of 14x. However, a large part of the algorithm remains, unaccelerated, at the host processor. This restricts the total speed-up to 2.2x. Predictions show that a total speed-up in the order of 8x should be achievable. The performance results are used to estimate the energy for Blocks using the energy model. These results show an energy improvement of 2.3x. Prediction indicates that energy improvement is restricted by the part of the algorithm still running at the host processor and that energy improvements around 5x should be achievable.

Although not part of this book, Blocks has been taped-out twice. Once as part of the CompSoC project where a small Blocks architecture was included as an accelerator to a ARM Cortex-M0 host processor. This small (4x4 function units) design was mainly intended as a functional verification. Evaluation has shown that this CGRA is functionally correct and can be used for small acceleration kernels. A chip photo of the CompSoC tape-out is shown in Fig. 8.1a. The underlying architecture of this system is shown in Fig. 8.1b. Blocks is shown in orange and marked as 'CGRA'.

A much larger architecture has been taped-out as part of the BrainWave project; this design is aimed at brain signal processing. Furthermore, a chip designed for

(a)

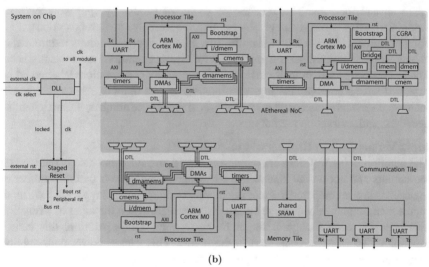

(b)

Fig. 8.1 Chip photo and architecture of the CompSoC tape-out. This tape-out included a Blocks instance that was shown functionally correct after tape-out. (**a**) Chip photo of the CompSoC tape-out including an instance of the Blocks processor (photo by Bart van Overbeeke). (**b**) System overview of the CompSoC tape-out, the Blocks instance is shown in orange and marked as 'CGRA' (image courtesy of the CompSoC team)

brain–computer interfaces, as described in Chap. 7, is under development. This design will have roughly double the number of function units as the BrainWave tape-out.

There is always a trade-off between energy efficiency and processor flexibility. Flexibility requires more possible configuration states in a processor which increases overhead. With Blocks a good balance between these two aspects can be achieved.

8.2 Future Work

A reconfigurable architecture such as Blocks provides a large number of possible optimizations and extensions. This section lists several of these extensions.

Switch-Box Network

The switch-boxes networks are an important part of the Blocks architecture. They are responsible for a significant part of the energy and area and as such are a prime target for further optimization. The switch-boxes that Blocks uses are multiplexer based and not optimal in power and area. Ongoing research shows that switch-box power and area can be reduced by over 40% and 20%, respectively, using pass-transistor designs. This reduces the power of Blocks by approximately 8%. Additional gains can be achieved by pruning the available switch-box connections without influencing routability significantly. Many filter applications require delay elements, for which function units that perform pass operations are currently used. The switch-boxes are, in their current version, entirely combinational, but could be extended with a configurable register on (some of) the outputs. Although this will increase the switch-box complexity, it allows to reduce the number of function units, decoders, and instruction memory.

Input Versus Output Register in Function Units

The function units of Blocks are output-registered, meaning that data is buffered on the outputs and not on the inputs. The advantage of doing so is that the number of registers is lower compared to input-registering as function units typically have more inputs than outputs. However, there is a disadvantage. When the input of a function unit changes, even though it might be performing a no-operation, this causes toggling of logic gates connected to the inputs. This causes power consumption that would be mitigated by placing the registers at the input of function units. The trade-off between these two options on real-life applications is not clear in the case of Blocks and should be further investigated.

Memory Hierarchy and Interface

The memory hierarchy is very important to achieve high performance on almost any processing platform. For Blocks, with its efficient use of parallelism that is available in algorithms, the importance of sufficient bandwidth is even more important. In the current design of the blocks memory interface the ciritical path runs from the LSUs, through the global memory arbiter, to the global data memory. Optimizing this interface is one of the remaining research questions to improve performance, and to allow better frequency scaling on sub-threshold-voltage technology libraries.

Besides critical path reduction, a more advanced memory hierarchy should be investigated. One primary point of investigation should be the inclusion of direct memory access (DMA) that can offload data management tasks from the Blocks fabric. Such a DMA should be capable of efficiently gathering data from memory before it is required and, equally importantly, allow access to this data in a manner that reduces processor stalls to a minimum. Coupled to such a system might be more advanced address generation in the load–store units, providing support for e.g. multi-dimensional convolutions. Although this would increase complexity, it may reduce the need for memory address computations on ALUs.

Optimizing Instruction Fetching

One aspect that has been partially investigated for Blocks is hardware loop support and instruction buffering. Although common for VLIW processors, implementing this within Blocks presents a few challenges. Hardware loop support and buffering has been shown to provide a significant reduction on instruction memory accesses, resulting in a lower energy. Another point of investigation is the use of instruction compression, also common on VLIW processors where the wide instructions cause a significant part of the energy consumption. Instruction compression on Blocks is not trivial since at design time it is not known which instruction decoders will operate in lock-step. Each of these instruction decoders has its own memory. To allow instruction compression multiple of these instruction memories need to be combined. This may be done at run-time with an extra stage in the instruction fetch pipeline, similar to the KAHRISMA architecture [1]. Alternatively, instruction memories may be clustered at design time, although this reduces the flexibility of Blocks, it opens up possibility for instruction compression and other energy saving optimizations.

Optimizing Reconfiguration Scheme

Configuration of the Blocks fabric is currently performed via a long scan-chain that passes through all building blocks that require configurations, typically these are function units and switch-boxes. The advantage of this is that it is very lightweight and allows the configuration bits to be exactly located where they are needed. Alternatively a memory based reconfiguration scheme could be considered. In such a scheme all configuration bits are arranged in a two-dimensional matrix which can be accessed like a typical memory, e.g. by memory word. Although this may lead to longer wires to the locations where the configuration is required, it allows faster configuration. More importantly, it allows updating only selected regions of the Blocks configuration, opening the way to partial reconfiguration.

Instruction Set Architecture Optimization

The current instruction set is based on the requirements of the benchmarks executed on Blocks and heavily influenced by those found on typical RISC-like processors. This selection of instructions works well for the current set of benchmark applications, but may not be optimal for other application domains. It might be interesting to investigate the energy impact of each of the instructions to the total design and how often they are actually used in a larger benchmark set. Additionally, not all instructions may be compiler friendly. Furthermore, additional instructions may make automatic compilation easier. Such specializations are very interesting for domains like deep learning. As a preliminary study, ternary neural network mapping has been investigated. With a limited number of specialized instructions, a large energy efficiency improvement can be accomplished for such domains. However, further investigation has to be performed to make the Blocks framework able to compete with specialized neural network architectures. Although Blocks will never be able to outperform a dedicated accelerator, it is flexible enough to be reused for accelerating other kernels. With this extension, Blocks may be more efficient for networks with layers of different precision.

Multi-Granular Function Units

The function units used in this book use a fixed width that is configured at design time. However, there is preliminary support from Blocks for multi-granular function units. This means that narrower units can be combined to form wider function units. For example, four 8-bit ALUs can be combined to form one 32-bit ALU. Similar options are available for multipliers. It is unclear however whether combining or splitting function units leads to more efficient implementations, this is something that should be investigated. Multi-granular function units are interesting because they allow adaptation to the data-width required for certain data-paths in an algorithm. For example, for algorithm operating mainly on 8-bit values, but computing some results with 32-bit wide data, it would reduce energy to only use a 32-bit data-path where needed and construct this out of 8-bit function units.

Additionally, multi-granular units could be used to incorporate approximate computing into the Blocks fabric. Special function units that provide approximate alternatives can be placed on the Blocks fabric. The routing configured onto the data network will then determine whether the computation will be exact or approximate. It is even possible to use two 8-bit exact ALUs to compute the higher 16-bits of a 32-bit result and two approximate units for the lower half. Of course, any mix between exact and approximate units could be configured. It is not required that function units are either exact or approximate. The configuration bit-stream could select whether the function unit operates in approximate or exact mode.

Multi-Processor Configuration

Blocks allows multi-processors to be instantiated onto its fabric. Such configurations likely require synchronization between the configured processors to perform certain operations at the right time. Section 3.7 already introduced some methods for doing so. However, this was never fully integrated and more, or more efficient, ways may be available.

Tool-Chain Improvements

For any architecture to be widely accepted, tool support is crucial. It is therefore important to automate the Blocks design flow as much as possible. The main component still lacking at this time is a compiler able to generate efficient PASM code from a higher level language. Generating code for Blocks is challenging as it does not fit very well in existing compiler frameworks such as LLVM. A collaborative effort is currently ongoing to adapt the TCE framework [2], suitable for TTA architectures, to allow code generation for Blocks.

Besides code generation, the Blocks framework requires a virtual architecture to be specified. It is, in theory, possible to extract such an architecture automatically from a high-level language description. Due to the large amount of design options, this tool will likely have to perform some design space exploration. Such a framework can make use of the architecture model described in Chap. 6.

Debugging is part of any design cycle. Currently the only debugging option for Blocks is to use an RTL-level simulator and debug the system by observing the signals. This is a rather slow process and requires getting used to. It would be better if a cycle accurate simulator would be implemented that can conveniently step through program to allow easier debugging. This may even be extended to hardware breakpoints such that actual hardware can be co-simulated.

References

1. R. Koenig, et al., KAHRISMA: A novel hypermorphic reconfigurable-instruction-set multi-grained-array architecture, in *2010 Design, Automation Test in Europe Conference Exhibition (DATE 2010)* (Mar. 2010), pp. 819–824. https://doi.org/10.1109/DATE.2010.5456939
2. P. Jääskeläinen, et al., HW/SW Co-design toolset for customization of exposed datapath processors, in *Computing Platforms for Software-Defined Radio* (Springer International Publishing, 2017), pp. 147–164. ISBN: 978-3-319-49679-5. https://doi.org/10.1007/9783319496795_8

Appendix A
The Blocks Function Units

The Blocks framework supports various function unit types. This appendix details the function units available in the Blocks framework. The currently available function units aim to provide basic signal processing capability to Blocks. However, the modular framework allows easy extension with new function units if required.

A.1 The Arithmetic Logic Unit (ALU)

The ALUs are the workhorse of the Blocks fabric. It supports operations for arithmetic and logic operations such as addition, subtraction, bit-wise operations, shifting, and comparisons. The comparisons that the ALU can perform allow it to generate conditions that are used in the control flow. The results of comparisons are stored both in an internal 'compare flag' register and on one of the ALUs outputs, as can be observed in Fig. A.1. The internal flag register can be used to control subsequent operations such as a conditional move. The compare result on the output can be the input for another function unit, such as the ABU. In Fig. A.1, the ALU is split into shifting logic and arithmetic sections. This is done to prevent toggling irrelevant logic where possible. For example, the result of a shift operation is not required to perform compare operations. The output of the logic section can be inverted to produce the negated version of all logic operations. For example, with one extra decoded instruction wire, the AND, OR, XOR, and pass operations can become NAND, NOR, XNOR, and negate operations, respectively. Inputs of the ALU can be sign-extended. Although it is possible to integrate this with the LSU, there is no space available in the instruction encoding for that unit.

Table A.1 shows all instructions that are supported by the ALUs. The notation is a modified version of the one used in the books by Patterson. Instead of specifying sources and destinations as registers, as common in the MIPS processor, the Blocks instructions specify inputs as sources and outputs as destinations. The inputs are

© The Author(s), under exclusive license to Springer Nature Switzerland AG 2022
M. Wijtvliet et al., *Blocks, Towards Energy-efficient, Coarse-grained Reconfigurable Architectures*, https://doi.org/10.1007/978-3-030-79774-4

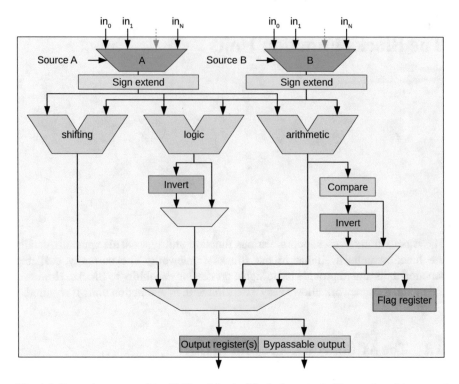

Fig. A.1 Internal structure of the ALU used for the Blocks framework. The number of inputs and outputs is a design parameter. The internal instruction connections are not shown for clarity

marked with I and the outputs with O. Which inputs or outputs are selected by the instruction is specified with inA, inB, and out. When sign extension is required this is denoted with the function $SE(...)$. If the data type is different from the width of the data-path, then this is specified with the $type$ argument. Results of compare operations are stored in r_f, which is the internal flag register, as well as made available on the outputs.

It can be observed that not all possible shift amounts are supported. This is done for energy reduction. An arbitrary shift is implemented as a large multiplexer array and therefore costly in area and energy. Instead, two shift amounts, that can be compounded to generate larger shift amounts, are provided. If arbitrary shift amounts are required it would be better to implement these in the multiplier unit as (part of) a divide operation.

Table A.1 Instructions for the ALU function unit. The flag register in Fig. A.1 is indicated with r_f in this table

Name	Mnemonic	Operation
No-operation	nop	
Addition	add	$O[out] = I[inA] + I[inB]$
Add. with sign-ext.	add_se	$O[out] = SE(I[inA], type) + SE(I[inB], type)$
Subtraction	sub	$O[out] = I[inA] - I[inB]$
Sub. with sign-ext.	sub_se	$O[out] = SE(I[inA], type) - SE(I[inB], type)$
And	and	$O[out] = I[inA] \& I[inB]$
Nand	nand	$O[out] =\sim (I[inA] \& I[inB])$
Or	or	$O[out] = I[inA] \mid I[inB]$
Nor	nor	$O[out] =\sim (I[inA] \mid I[inB])$
Xor	xor	$O[out] = I[inA] \wedge I[inB]$
Xnor	xnor	$O[out] =\sim (I[inA] \wedge I[inB])$
Negate	neg	$O[out] =\sim (I[inA])$
Conditional move	cmov	$O[out] = r_f \ ? \ I[inA] \ : \ I[inB]$
External cond. move	ecmov	$O[out] = I[0] \ ? \ I[inA] \ : \ I[inB]$
Pass	pass	$O[out] = I[inA]$
Pass with sign-ext.	pass_se	$O[out] = SE(I[inA], type)$
Test equal	eq	$r_f = (I[inA] == I[inB]) \ ? \ (1 \ : \ 0) \ ; \ O[out] = r_f$
Test not equal	neq	$r_f = (I[inA]! = I[inB]) \ ? \ (1 \ : \ 0) \ ; \ O[out] = r_f$
Test less (unsigned)	ltu	$r_f = (I[inA] < I[inB]) \ ? \ (1 \ : \ 0) \ ; \ O[out] = r_f$
Test less (signed)	lts	$r_f = (SE(I[inA]) < SE(I[inB])) \ ? \ (1 \ : \ 0) \ ; \ O[out] = r_f$
Test greater or equal (unsigned)	geu	$r_f = (I[inA] >= I[inB]) \ ? \ (1 \ : \ 0) \ ; \ O[out] = r_f$
Test greater or equal (signed)	ges	$r_f = (SE(I[inA]) >= SE(I[inB])) \ ? \ (1 \ : \ 0) \ ; \ O[out] = r_f$
Shift logic left by 1	shll1	$O[out] = I[inA] << 1$
Shift logic left by 4	shll4	$O[out] = I[inA] << 4$
Shift logic right by 1	shrl1	$O[out] = I[inA] >> 1$
Shift logic right by 4	shrl4	$O[out] = I[inA] >> 4$
Shift arithmetic right by 1	shra1	$O[out] = SE(I[inA] >> 1)$
Shift arithmetic right by 4	shra4	$O[out] = SE(I[inA] >> 4)$

A.2   The Accumulate and Branch Unit (ABU)

As mentioned in Sect. 3.1, one of the function unit types in the Blocks fabric provides sequencing for the control flow. This function supports program counter computation based on common jump and branch operations. Blocks must provide both predetermined control flow as data dependent control flow. Predetermined control flow is based on static loop bounds and can be specified as fixed addresses where jumps occur until the required number of loop iterations is completed. Data dependent flow control cannot be determined in advance and requires conditions to be evaluated during program flow. This requires blocks to support branches that evaluate conditions that are calculated inside the data-path and determine whether to branch based on those conditions.

The Blocks ABU supports both of these kinds of branching. Static schedules can either be implemented with branch instructions computed by other function units, or by using hardware loop support in the ABU.[1] The hardware loop support uses a loop starting point, loop end point, and a loop count to configure a hardware loop. At every loop end the loop counter is evaluated, depending on its value the program counter is either set to the loop start point or incremented by one. The hardware loop support uses three registers for each loop level. The ABU has a configurable number of register entries (by default 16), in its register file, that can be used to store hardware loop configurations. This allows for four (nested) loops to be stored by default. Hardware loops are configured by writing to the corresponding registers in the register file. This is typically done just before the loops are required and is performed by the application.

Data dependent branching is supported by the branch instructions. These instructions can be (un)conditional branches, both relative and absolute. Conditional branches can be evaluated based on an internal loop bound computation (static loop support), or based on an external condition connected to one of the data network inputs of the ABU. Table A.2 shows the instructions supported by the ABU when performing branches.

There can be multiple ABUs present within the same Blocks fabric, allowing for multi-processor configurations. However, most applications do not require such configurations. To prevent having function units on the fabric that are only useful in specific cases, the ABU does not only support branching but also accumulation. When the operations required to support branches are investigated it can be observed that computing a program counter and performing accumulation are very similar. This implies that the underlying hardware required for branching can be reused for accumulation. The accumulation results are stored in the register file. Depending on the number of configured registers it is possible to store multiple accumulation results. The functionality of the ABU is configured using a bit in the bit-stream and is therefore static during application execution. Figure A.2a shows the structure of the ABU when configured for branch mode, Fig. A.2b shows the ABU in accumulation mode. Table A.3 shows the supported ABU instructions in accumulation mode. Only the highest output number of the ABU can select a register as a source, the lower outputs are directly connected to the corresponding

---

[1]Hardware loop support was implemented by Kanishkan Vadivel as part of his master thesis project [1].

Table A.2 Instructions for the ABU function unit in branch mode

Name	Mnemonic	Operation
No-operation	nop	$PC = PC + 1$
Jump relative	jr	$PC = PC + SE(I[inB])$
Jump absolute	ja	$PC = I[inB]$
Branch conditional relative	bcr	$PC = (I[inA]! = 0) ? PC + SE(I[inB]) :$ $PC + 1$
Branch conditional absolute	bca	$PC = (I[inA]! = 0) ? I[inB] : PC + 1$
Jump relative immediate	jri	$PC = PC + SE(immediate)$
Jump absolute immediate	jai	$PC = immediate$
Branch conditional relative immediate	bcri	$PC = (I[inA]! = 0) ? PC + SE(immediate) : PC + 1$
Branch conditional absolute immediate	bcai	$PC = (I[inA]! = 0) ? immediate : PC + 1$
Store register immediate	srm	$R[immediate] = I[inA]$
Load register immediate	lrm	$O[N - 1] = R[immediate]$

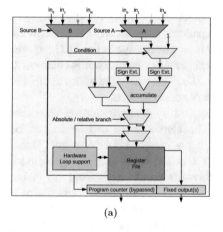 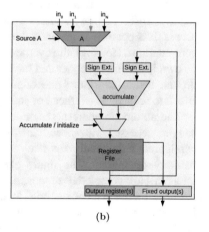

(a)   (b)

Fig. A.2 The accumulate and branch unit of Blocks. (**a**) shows the branching configuration while (**b**) shows the accumulation configuration. (**a**) ABU configuration for branching. (**b**) ABU configuration for accumulation

register number. For example, output 0 is directly connected to register 0 in the register file.

A.3   The Load–Store Unit (LSU)

The LSU is the interface to the memory hierarchy for all other function units on the Blocks fabric. The Blocks LSUs support byte (8-bit), half-word (16-bit), and word

Table A.3 Instructions for the ABU function unit in accumulate mode

Name	Mnemonic	Operation
No-operation	nop	
Accumulate unsigned	accu	$R[immediate] =$ $R[immediate] + I[inA]$
Accumulate signed	accu	$R[immediate] =$ $R[immediate] + SE(I[inA])$
Store register immediate	srm	$R[immediate] = I[inA]$
Load register immediate	lrm	$O[N-1] = R[immediate]$

(32-bit) operations on both the local and global memory. Local memory operations are performed on memories that are private to the LSU and can therefore not incur stalls. Global memory operations are performed through an arbiter and can be stalled when contention on the memory interface occurs.

For many digital signal processing applications the memory access patterns can be very predictable. In this case, the access patterns can often be described as offsets relative to the memory address space and increments (stride) for each memory access. It is possible to generate these access patterns using an ALU to perform the computations. However, the overhead for this is relatively large as the ALU requires its own instruction decoder and the associated instruction memory. It is common to provide dedicated address generators within LSUs. This technique is also applied in Block to reduce the overhead of strided memory access patterns.

The automatic address generation is controlled by configuration registers that can be accessed via special instructions. The start address and stride can be configured individually for read and write operations and for local and global memory operations. The figure also shows that the local and global memory interfaces are completely separated. This allows for simultaneous local and global memory operations. For example, loading a value from global memory and storing it in local memory. The same holds for read and write operations to the same memory. As long as the memory connected to the interface allows parallel read and write operations, the LSU does support this. If this is not supported by the memory, then the LSU will perform the read and write accesses sequentially. Doing so makes the system very independent from the type of memory (or memory bus) connected to the LSU. Figure A.3 shows the internal structure of the LSU. All instructions supported by the LSU, including parallel memory operations, are shown in Table A.4.

A.4   The Multiplier Unit (MUL)

The multiplier can perform, as the name suggests, multiplication operations. These multiplications can be both signed and unsigned and are performed on fixed point data representations. In fixed point notations the application developer chooses where the radix point is located. The radix point is the position below which

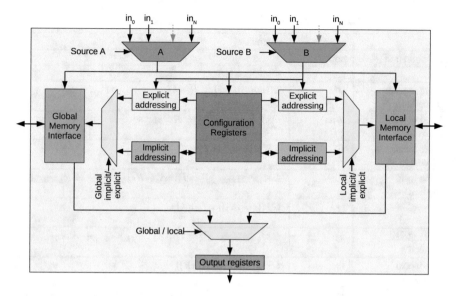

Fig. A.3 Internal structure of the LSU used for the Blocks framework

each binary bit represents a value below one. For example, if we place the radix point in the binary sequence 01011010 between the fifth and fourth bit (and obtain 0101.1010) this binary sequence represents the decimal value 5.625. The location of the radix point does not change the way that the hardware computes results, it is merely a decision on what the number represents. The notation for fixed point numbers is usually in the form $Qm.f$ where m represents the magnitude and f represents the fraction. For example, $Q23.8$ represents a 32-bit binary format where the higher 24 bits are used for the magnitude and the lower 8 bits are used for the fraction. A sign bit is implied, forming a complete 32-bit word.

When two 32-bit fixed point numbers are multiplied the result is a 64-bit number. Additionally, the radix point moves. For example, multiplying two numbers in Q23.8 format yields a Q46.16 fixed point number. To continue computation using Q23.8 numbers the multiplication first has to be normalized. In the example this can be achieved by shifting the result right by eight bits.

By supporting 8, 16, and 24 bits shifting operations after multiplication results can directly be normalized. Figure A.4 shows the internal structure of the multiplier. When a multiplication is performed, the data-width of the result is twice as large as the data-width of the input data. Therefore, the multiplier in Blocks has a special register where the upper half of the result can be stored and read-back separately. It is also possible to use two outputs onto the network to communicate the result at once. Table A.5 describes the instructions supported by the multiplier unit.

Table A.4 Instructions for the LSU function unit

Name	Mnemonic	Operation
No-operation	nop	
Pass	pass	$O[out] = I[inA]$
Store register immediate	srm	$R[immediate] = I[inA]$
Load register immediate	lrm	$O[N-1] = R[immediate]$
Store local addressed	sla	$LM[I[inB]] = I[inA]$
Store local implicit	sli	$LM[R_{LMaddr\_S}]] = I[inA]$; $R_{LMaddr\_S}+ = R_{LMstride}$
Store global addressed	sga	$GM[I[inB]] = I[inA]$
Store global implicit	sgi	$GM[R_{GMaddr\_S}]] = I[inA]$; $R_{GMaddr\_S}+ = R_{GMstride}$
Load local addressed	lla	$O[out] = LM[I[inB]]$
Load local addressed	lli	$O[out] = LM[R_{GMaddr\_L}]]$; $R_{GMaddr\_L}+ = R_{GMstride}$
Load local addressed	lga	$O[out] = GM[I[inB]]$
Load local addressed	lgi	$O[out] = GM[R_{GMaddr\_L}]]$; $R_{GMaddr\_L}+ = R_{GMstride}$
Parallel lli and sla	lli_sla	$lli(out) \parallel sla(inB, inA)$
Parallel lgi and sla	lgi_sla	$lgi(out) \parallel sla(inB, inA)$
Parallel lli and sga	lli_sga	$lli(out) \parallel sga(inB, inA)$
Parallel lgi and sga	lgi_sga	$lgi(out) \parallel sga(inB, inA)$
Parallel lli and sli	lli_sli	$lli(out) \parallel sli(inA)$
Parallel lgi and sli	lgi_sli	$lgi(out) \parallel sli(inA)$
Parallel lli and sgi	lli_sgi	$lli(out) \parallel sgi(inA)$
Parallel lgi and sgi	lgi_sgi	$lgi(out) \parallel sgi(inA)$
Parallel lla and sli	lli_sli	$lla(out, inB) \parallel sli(inA)$
Parallel lga and sli	lli_sli	$lga(out, inB) \parallel sli(inA)$
Parallel lla and sgi	lli_sli	$lla(out, inB) \parallel sgi(inA)$
Parallel lga and sgi	lli_sli	$lga(out, inB) \parallel sgi(inA)$

A.5   The Register File (RF)

In most processors register file read and write operations are to a certain extent implicit; there is no explicit memory read command when, for example, an addition has to be performed.

Table A.5 Instructions for the MUL function unit

Name	Mnemonic	Operation
No-operation	nop	
Pass	pass	$O[out] = I[inA]$
Unsigned mult., output lower half	mullu	$\{R_h, O[out]\} = I[inA] \cdot I[inB]$
Unsigned mult., output lower half shifted by 8	mullu_sh8	$\{R_h, O[out]\} = (I[inA] \cdot I[inB]) >> 8$
Unsigned mult., output lower half shifted by 16	mullu_sh16	$\{R_h, O[out]\} = (I[inA] \cdot I[inB]) >> 16$
Unsigned mult., output lower half shifted by 24	mullu_sh24	$\{R_h, O[out]\} = (I[inA] \cdot I[inB]) >> 24$
Signed mult., output lower half	mulls	$\{R_h, O[out]\} =$ $SE(I[inA]) \cdot SE(I[inB])$
Signed mult., output lower half shifted by 8	mulls_sh8	$\{R_h, O[out]\} =$ $SE((SE(I[inA]) \cdot SE(I[inB])) >> 8)$
Signed mult., output lower half shifted by 16	mulls_sh16	$\{R_h, O[out]\} =$ $SE(SE(I[inA]) \cdot SE(I[inB])) >> 16)$
Signed mult., output lower half shifted by 24	mulls_sh24	$\{R_h, O[out]\} =$ $SE(SE(I[inA]) \cdot SE(I[inB])) >> 24)$
Unsigned multiplication	mulu	$O[out + 1, out] = I[inA] \cdot I[inB]$
Unsigned multiplication, shifted by 8	mulu_sh8	$O[out+1, out] = (I[inA] \cdot I[inB]) >> 8$
Unsigned multiplication, shifted by 16	mulu_sh16	$O[out + 1, out] = (I[inA] \cdot I[inB]) >> 16$
Unsigned multiplication, shifted by 24	mulu_sh24	$O[out + 1, out] = (I[inA] \cdot I[inB]) >> 24$
Signed multiplication	muls	$O[out + 1, out] =$ $SE(I[inA]) \cdot SE(I[inB])$
Signed multiplication, shifted by 8	muls_sh8	$O[out + 1, out] =$ $SE((SE(I[inA]) \cdot SE(I[inB])) >> 8)$
Signed multiplication, shifted by 16	muls_sh16	$O[out + 1, out] =$ $SE(SE(I[inA]) \cdot SE(I[inB])) >> 16)$
Signed multiplication, shifted by 24	muls_sh24	$O[out + 1, out] =$ $SE(SE(I[inA]) \cdot SE(I[inB])) >> 24)$
Load higher result register	lh	$O[out] = R_h$

Table A.6 Instructions for the RF function unit

Name	Mnemonic	Operation
No-operation	nop	
Store register immediate	srm	$R[immediate] = I[inA]$
Load register immediate	lrm	$O[N - 1] = R[immediate]$
Store register addressed	sra	$R[I[inB]] = I[inA]$
Load register addressed	lra	$O[N - 1] = R[I[inB]]$
Parallel lrm and srm	lrm_srm	$lrm(immediate_L) \parallel srm(immediate_R, inA)$

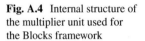

Fig. A.4 Internal structure of
the multiplier unit used for
the Blocks framework

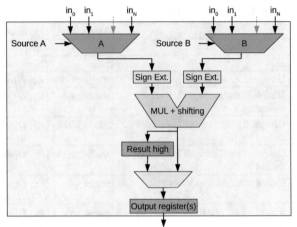

Register files in Blocks are separate function units. Like other function units in Blocks, the inputs and outputs of a register file are connected to other function units over the data network. A register file is controlled by an instruction decoder that issues instructions to the register file. This instruction decoder has no connection to other instruction decoders and therefore has no knowledge of what instructions are performed by other function units. For this reason, the instruction executed by, for example, an ALU cannot control operation of the RF. Therefore, the instruction decoder of the RF has to be explicitly instructed to perform a read or write operation to one of its registers. Registers operations in Blocks are therefore explicit.

The registers can be read and the results can be made available on a specified output port as well as stored into a register location from a specified input port. The register file supports simultaneous register reads and writes as shown in the instruction specification in Table A.6.

The multiplexer that allows selection of which register is being read gives both energy and area overhead and requires space in the instruction format. The number of registers in the register file in Blocks is a design parameter (default 16). The default number of registers requires four bits in the instruction format to configure the output multiplexer. For multiple selectable output ports multiple multiplexers would be required, each requiring their own bits in the instruction format. By using bypassing between function units the register pressure can be significantly reduced in most DSP applications, therefore usually only requiring one selectable output port per register file. For this reason the Blocks register files only have one selectable output. But in addition, there are fixed outputs that are connected to fixed register locations, as can be seen in Fig. A.5. For example, output 0 is always connected to register location 0, thereby not requiring a multiplexer or space in the instruction format. This way some working registers can be used without taking up space in the instruction encoding or incurring energy overhead by requiring multiplexers.

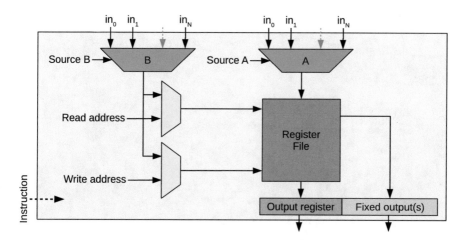

Fig. A.5 Internal structure of the register file used within the Blocks framework

Table A.7 Instructions for the IU function unit

Name	Mnemonic	Operation
No-operation	nop	
Load immediate	imm	$O = immediate$

A.6 The Immediate Unit (IU)

Although constant values can be read from the global memory, these extra memory accesses both reduce performance and energy efficiency. This means that all required constants should be stored in global memory such that they can be read using the LSUs. These memory accesses can be avoided by the use of immediate units. An immediate unit is able to produce constants on the data network. These constants are used for various purposes such as configuring registers at application initialization and providing constants during operations. The width of the instructions required for the immediate unit differs from the instructions for other function units. Therefore, the immediate unit contains its own instruction decoder. Table A.7 shows the instructions for the immediate unit. It is possible to construct, and use, architecture instances that do not have an immediate unit.

Reference

1. K. Vadivel, et al., Loop overhead reduction techniques for coarse grained reconfigurable architectures, in *DSD 2017 - 20th Euromicro Conference on Digital System Design, Vienna* (2017)

Index

© The Author(s), under exclusive license to Springer Nature Switzerland AG 2022
M. Wijtvliet et al., *Blocks, Towards Energy-efficient, Coarse-grained Reconfigurable
Architectures*, https://doi.org/10.1007/978-3-030-79774-4